Elements of Radio Frequency Energy Harvesting and Wireless Power Transfer Systems

Elements of Radio Frequency Energy Harvesting and Wireless Power Transfer Systems

Taimoor Khan, Nasimuddin, and
Yahia M.M. Antar

CRC Press is an imprint of the
Taylor & Francis Group, an **informa** business

First edition published 2021
by CRC Press
6000 Broken Sound Parkway NW, Suite 300, Boca Raton, FL 33487-2742

and by CRC Press
2 Park Square, Milton Park, Abingdon, Oxon, OX14 4RN

First edition published by CRC Press 2021

CRC Press is an imprint of Taylor & Francis Group, LLC

ISBN: 9780367246785 (hbk)
ISBN: 9780429283918 (ebk)

Typeset in Times
by codeMantra

Contents

PART A Elements of RF Energy Harvesting (RFEH) Systems

List of Figures

List of Tables

Preface

Revolution in the wireless technology has created an Information Technology Society that has changed the way we communicate. Demands for mobile services are increasing, and new applications are emerging; as a result, developing new wireless systems has become a critical and fundamental issue. In the next decade, it is expected that the performance of future wireless communication technologies will need to improve by up to 1000x over existing systems to meet the demands. With the advances and popularity of wireless communication devices, a large amount of abundant radio frequency (RF) energy from surrounding sources is being scattered in the environment. This scattered RF energy is also creating another problem, called electromagnetic (EM) pollution, in equal proportion. The EM pollution is harmful for both the living and non-living things in the environment. Using an efficient RF harvester, these EM waves can be recycled and converted into electrical energy. This method is called *energy harvesting*, also known as *energy scavenging*, in which the surrounding EM waves are converted into electrical energy. An RF energy harvesting system basically consists of a receiving antenna, matching, and a rectifying circuit. The receiving antenna receives EM waves from the surrounding environment and then converts them into AC voltage and current. The rectifying circuit converts the AC power generated from the input EM waves into DC power. The RF energy harvesting antennas are instigated in the receiver side to receive the EM waves from the various broadly available sources such as GSM (0.9 and 1.8 GHz), UMTS (1.92–2.17 GHz), LTE (2.3–2.4 GHz), and WLAN/WiMAX/Wi-Fi (2.4, 3.6, 4.9, 5, and 5.9 GHz bands). In several areas, such as low-power wireless sensors, radio frequency identification (RFID) tags, and biotelemetry, the demand of an alternate power source has increased a lot in last couple of years. Periodic battery replacements for a large number of nodes are unrealistic and expensive. Hence, scavenging of ambient RF energy has gained massive popularity in the literature. However, the efficiency of energy harvesting needs to be improved, and creative ways for achieving this still need to be found. The efficiency of an energy harvesting system depends on the efficiency of each component (antenna circuit, rectifier circuit, rectenna circuit, power management circuit, etc.) used in it. The proposed book is based on the theme to compile a comprehensive volume on the research carried out on harvester elements (antenna circuits, rectifier circuits, impedance matching circuits, or rectenna circuits), which will be very much helpful in developing more efficient RF energy harvesters based on ambient RF energy from the abovementioned sources.

Wireless power transfer (WPT), wireless power transmission, wireless energy transmission, or EM power transfer is the transmission of electrical energy without wires as a physical link. In a WPT system, a transmitter device, driven by electrical power from a power source, generates a time-varying EM field, which transmits power across space to a receiver device, which extracts power from the field and supplies it to an electrical load. WPT is useful to supply power to electrical devices where interconnecting wires are inconvenient, hazardous, or not possible. WPT techniques mainly fall into two categories: non-radiative and radiative. In *near-field* or *non-radiative* techniques, power is transferred over short distances either by creating magnetic fields using inductive coupling between coils of wire or by creating electric fields using capacitive coupling between metal electrodes. Inductive coupling is the most widely used wireless technology; its applications include charging handheld devices such as mobile phones and electric toothbrushes, RFID tags, and wirelessly charging or continuously transferring wireless power in implantable medical devices like artificial cardiac pacemakers, or electrical vehicles. In *far-field* or *radiative* techniques, also called *power beaming*, power is transferred by beams of EM radiation, such as microwaves or laser beams. These techniques can transport energy to longer distances, but the transmitter must be aimed at the receiver. The proposed applications of this type are solar power satellites and wireless-powered drone aircraft. An important issue associated with all wireless power systems is limiting the exposure of people and other living things to potentially injurious EM fields. WPT is a generic term for a number of different technologies for transmitting energy by means of EM technologies, but these technologies differ based on the following: the distance over which they can transfer power efficiently; whether the transmitter must be aimed (directed) at the receiver; and the type of EM energy they use such as time varying electric fields, magnetic fields, radio waves, microwaves, infrared or visible light waves. In general, a wireless power system consists of a *transmitter* device connected to a source of power such as a mains power line, which converts the power to a time-varying EM field, and one or more *receiver* devices, which receive the power and convert it back to DC or AC electrical current which is used by an electrical load. At the transmitter, the input power is converted to an oscillating EM field by some type of *antenna* device. The word *antenna* is used loosely here; it may be a coil of wire that generates a magnetic field, a metal plate that generates an electric field, an antenna that radiates radio waves, or a laser that generates light. A similar antenna or coupling device at the receiver converts the oscillating fields to an electric current. An important parameter that determines the type of waves is their frequency. Wireless power uses the same fields and waves that are used in wireless communication devices such as radio and television, both of which are familiar technologies that involve electrical energy transmitted without wires by EM fields; cell phones; and Wi-Fi routers. In radio communication, the goal is the transmission of information, so the amount of power reaching the receiver is not so important, as long as it is sufficient that the information can be received intelligibly. In wireless communication technologies, only small amounts of power reach the receiver. In contrast, with wireless power, the amount of energy received is the most important thing, so the efficiency (fraction of transmitted energy that is received) is the most significant parameter. For this reason, wireless power technologies are likely to be more limited by distance than wireless communication technologies. WPT may be used to power up wireless information transmitters or receivers. This type of communication is known as wireless-powered communication (WPC). When the harvested power is used to supply the power of wireless information transmitters, the network is known as Simultaneous Wireless Information and Power Transfer (SWIPT), whereas when it is used to supply the power of wireless information receivers, it is known as a

Wireless-Powered Communication Network (WPCN). In the United States, the Federal Communications Commission (FCC) provided its first certification for a wireless transmission charging system in December 2017.

Our main objective was to provide a comprehensive book that consists of comprehensive descriptions of different elements and the emerging areas of RF energy harvesting and wireless power transfer systems.

Taimoor Khan
Nasimuddin
Yahia M.M. Antar

Acknowledgments

Taimoor Khan and Yahia M.M. Antar acknowledge the support received from the Ministry of Human Resource Development (MHRD), Government of India under the Scheme for Promotion of Academic and Research Collaboration (SPARC) (Grant No. SPARC/2018–2019/P266/SL/2019).

Authors

Taimoor Khan is Assistant Professor in the Department of Electronics and Communication Engineering at National Institute of Technology Silchar, Assam since 2014. Prior to joining NIT Silchar, he served several organizations viz. Netaji Subhas Institute of Technology Patna, as an Associate Professor and Head of the Electronics and Communication Engineering Department for more than a year; Delhi Technological University (Formerly Delhi College of Engineering), Govt. of NCT of Delhi, Delhi, as an Assistant Professor for more than two years; and Shobhit Institute of Engineering and Technology (A Deemed to be University), Meerut, as a Lab Instructor, Lecturer and Assistant Professor for more than nine years. In addition to this regular experience, he has also worked as Visiting Researcher at Queen's University Canada and Royal Military College of Canada under ongoing international collaborative SPARC project during September-October 2019. Before that, Dr. Khan has also worked as a Visiting Assistant Professor at Asian Institute of Technology, Bangkok, Thailand under a short term deputation programme of Ministry of the HRD, Govt. of India during September-December 2016. Dr. Khan was awarded his Ph.D. Degree in Electronics and Communication Engineering from National Institute of Technology Patna, in 2014; Master Degree in Communication Engineering from Shobhit Institute of Engineering and Technology (A Deemed to be University), Meerut in 2009; Bachelor Degree in Electronics and Communication Engineering from The Institution of Engineers (India), Kolkata, in the year 2005 and Polytechnic Diploma in Electronics Engineering from Government Polytechnic Saharanpur, Uttar Pradesh in 2001. His active research interest includes Printed Microwave Circuits, Electromagnetic Bandgap Structures, Ultrawideband Antennas, Dielectric Resonator Antennas, Ambient Microwave Energy Harvesting and Artificial Intelligence Paradigms in Electromagnetics. Dr. Khan has successfully guided two Ph.D. Theses in the area of antenna engineering and allied domain and third one was submitted in August 2020. Presently he is guiding four PhD students, in the department of Electronics and Communication Engineering at NIT Silchar. He has published over seventy research papers in SCI/SCIE journals of repute like; IEEE Magazine, IEEE Letters, IET Microwaves, Antennas and Propagation, Progress in Electromagnetic research, etc. as well as in world reputed conference proceedings including URSI RCRS, URSI GASS, IEEE European Microwave Week, IEEE AEMC, IEEE APACE, etc. Currently, Dr. Khan is executing three funded research projects including one international collaborative research project with Queen's University Canada under the SPARC Scheme of Ministry of the HRD, Govt. of India. In July 2020, he was granted another international collaborative research project with California State University, Northridge, USA under SERB-VAJRA Scheme of Govt. of India. Dr. Khan is an active Senior Member of IEEE (USA), Senior Member of URSI (Belgium) and Fellow of IETE (India). He is also a member of the Editorial Board of Wiley's International Journal of RF and Microwave Computer Aided Engineering.

Nasimuddin is a scientist at the Institute for Infocomm Research (I2R), Agency for Science, Technology and Research (A*STAR), Singapore. He received his Bachelor's Degree (Physics, Chemistry and Maths) in 1994 from Jamia Millia Islamia, New Delhi, India, and Master's Degree (Microwave Electronics) and Ph.D. Degree (Electronic Science) in 1998 and 2004, respectively, from the University of Delhi, India. He has worked as a Senior Research Fellow (1999–2003) in a DST-sponsored project on 'Optical Control of Passive Microwave Devices' in Council of Scientific and Industrial Research (CSIR), Government of India, and as a Senior Research Fellow in Engineering Science for the project "Investigations of microstrip antennas as a sensor for determination of complex dielectric constant of materials" at Department of Electronic Science, University of Delhi, India. He has worked as an Australian Postdoctoral Research Fellow (2004–2006), receiving the Discovery project grant from Australian Research Council, for the project "Microwave sensor based on multilayered microstrip patch/line resonators" at the Macquarie University, Australia. He has published 190 journal and conference technical papers on microstrip-based microwave antennas and components. He has edited and contributed a chapter to the book *Microstrip Antennas* published in 2011 by InTech. His research interests include multilayered microstrip-based structures, millimeter-wave antennas, radiofrequency identification reader antennas, Global Positioning System/Global Navigation Satellite System, ultra-wideband antennas, metamaterials-based microstrip antennas, satellite antennas, RF energy harvesting systems, circularly polarized microstrip antennas, and small antennas for TV white space communications. He is a senior member of the IEEE and the IEEE Antennas and Propagation Society and an executive committee member of IEEE MTT/AP chapter/CRFID chapter Singapore. He has been an organizing committee member of international/national conferences related to antenna and propagation. He was awarded a senior research fellowship from the Council of Scientific and Industrial Research, Government of India in Engineering Science (2001–2003); a Discovery Projects fellowship from the Australian Research Council (2004–2006); Singapore Manufacturing Federation Award (with project team) in 2014; and the Young Scientist Award from the URSI in 2005.

Yahia M.M. Antar joined the Department of Electrical and Computer Engineering at the Royal Military College of Canada, Kingston in November 1987 where he has held the position of Professor since 1990. He received his Bacheor's Degree in 1966 in B.Sc. (Hons.) Electrical Engineering (Communications) from Alexandria University. 1971 M.Sc. Electrical Engineering, University of Manitoba. 1975 Ph.D. Electrical Engineering, University of Manitoba. He has authored and co-authored more than 200 journal papers, several books and chapters in books, and more than 450 refereed conference papers, and holds several patents. He has supervised and co-supervised more than 90 Ph.D. and M.Sc. theses at the Royal Military College and at Queen's University. Dr. Antar is a life fellow of the IEEE, a fellow of the Engineering Institute of Canada (FEIC), a fellow of the Electromagnetic Academy, and a fellow of URSI. He served as the Chair of CNC, URSI (1999–2008), Commission B (1993–1999), and also as an Associate Editor of many IEEE and IET journals and as an IEEE-APS Distinguished Lecturer. In May 2002, he was awarded a Tier 1 Canada Research Chair in Electromagnetic Engineering which has been renewed in 2016. In 2003, he was awarded the Royal Military College of Canada "Excellence in Research" Prize, and the RMCC Class of 1965 Teaching Excellence Award in 2012. He was elected by the URSI Board as Vice President in 2008 and in 2014, and to the IEEE AP AdCom. in 2009. On January 31, 2011, he was appointed as a Member of the Canadian Defence Advisory Board (DAB) of the Canadian Department of National Defence. In October 2012, he received the Queen's Diamond Jubilee Medal from the Governor General of Canada in recognition of his contribution to Canada. He is the recipient of the 2014 IEEE Canada RA Fessenden Silver Medal for "Ground Breaking Contributions to Electromagnetics and Communications" and the 2015 IEEE Canada J. M. Ham outstanding Engineering Education Award. In May 2015, he received the Royal Military College of Canada Cowan Prize for excellence in research. He is the recipient of the IEEE-AP-S the Chen-To-Tai Distinguished Educator Award for 2017. In 2019, Prof. Antar was elected as the President Elect from Regions 7–10 (One-Year Term, January 1, 2020–December 31, 2020).

Part A

Elements of RF Energy Harvesting (RFEH) Systems

1 Introduction

1.1 INTRODUCTION

The ever-increasing global population and its increasing energy consumption raise the demand to adopt different approaches for utilizing the available renewable or green energy sources. This is necessary for efficiently utilizing the environment and partially fulfilling the global demand, as well as reducing environmental pollution. Because of these advantages, energy harvesting (EH) approaches have gained more popularity in recent years.

The EH is the process of capturing the energy available in the surrounding environment of a system and converting it into a suitable electrical form for supplying power to low-power electronic devices. Thus, the EH technique could be a better alternative for systems operating through wired networks. Different applications in sensing and implantable devices require batteries or wired cables for powering themselves, which sometimes makes their operation very difficult and complicated. EH approaches are more useful in the processes where batteries need to be periodically replaced in devices mounted in remote places like hill stations, etc. Thus, the lifetime of these devices also increases due to the EH strategies, and thereby reducing the device maintenance cost.

EH strategies are of two types: the first one is the ambient EH approach, and the second one is a dedicated source of approach. In the first approach, the harvesting system utilizes the green energy from the accessible sources in the surrounding environment of a system, and in the second approach, a dedicated source of energy is established to deliver the required amount of energy for consumer devices over the free space without causing any environmental pollution. Based on the first approach, several EH methods are existing, that harvest energy from never-ending sustainable energy sources such as thermal, wind, photo-voltaic (PV), vibrations, and electromagnetic (EM) waves [1–28]. These sources of energy remain available abundantly, and it can also be collected from the ambient environment.

1.2 VARIOUS ENERGY HARVESTING TECHNIQUES

1.2.1 Photo-Voltaic Energy Harvesting

In this process, the incoming photons from the solar or other light sources are converted into electricity. Here, PV cells are used for harvesting photons. This is a traditional and commercially reputable harvesting methodology. This technology delivers higher levels of output power than any other EH technologies. However, the harvesting ability of the system primarily depends on the intensity of light present and also on environmental conditions. A special type of setup is needed for harvesting the energy, and it cannot be operated in the indoor environment. This is the main limitation of the PV EH system [7–12].

1.2.2 Thermal Energy Harvesting

In this approach, the electricity is produced based on the temperature variation or gradient. As long as there is a difference in temperature, the system produces electricity. Otherwise, the produced voltage disappears due to current leakage. This form of energy is available only in larger units, which is the primary difficulty in the conversion of this energy to get sufficient output power from this harvesting system. Moreover, harvesting devices are larger in size as compared to all other EH techniques. Thus, a large area is required to deliver an adequate amount of output power from this thermal EH system [7,8,13–17].

1.2.3 Mechanical Energy Harvesting

In this process, electricity is produced from mechanical stress, pressure, and vibrations. The piezoelectric generators used for mechanical EH are lightweight and quite small in size. However, the output power depends on the degree of motion/vibration of the object, and the output power is highly inconsistent when motion/vibration is unstable. Thus, the conversion efficiency is very less due to variation in input power [7,8,13,18].

1.2.4 Wind Energy Harvesting

In this approach, the amount of output power delivered depends on the environmental conditions. It is highly variable and unpredictable; hence, the output power is low with poor conversion efficiency performance [7,8,19,20].

1.2.5 Acoustic Energy Harvesting

In this process, an acoustic transducer is used to convert continuous acoustic waves into electricity. This is useful where electrical communications are not possible. Generally, the available acoustic energy is low in density, and it is collected especially in a noisy environment. Thus, the harvested power by the acoustic transducer is very poor in quantity, and generated output power is very low as compared to all other EH techniques [21–23].

1.2.6 Electromagnetic (EM)/Radio Frequency (RF) Energy Harvesting

In this technique, electromagnetic energies available in the ambient environment are harvested and converted into electrical energy. However, densities of the available electromagnetic energies are very low in the surrounding environment, but the energy is continuous irrespective of diurnal and seasonal variations. The main advantage of this EH technology is that it can be operated efficiently even in indoor environments. The space occupied by the harvesting system is small in dimension. A special type of device, known as a rectenna or a rectifying device, is used for capturing electromagnetic energy [7,8,13,24–27].

TABLE 1.1
Several EH Techniques and their Performances

Energy Source	Available Input Power Density	Output Power	Availability
Ambient light:			
Indoor	0.1 mW/cm^2	10 μW/cm^2	During day
Outdoor	100 mW/cm^2	10 mW/cm^2	
Vibration/motion:			
Human	0.5 m/s^2 @ 1 Hz, 1 m/s^2 @ 50 Hz	4 μW/cm^2	Depending on the movement
Industrial	1 m/s^2 @ 1 Hz, 10 m/s^2 @ 1 KHz	100 μW/cm^2	
Thermal:			
Human	20 mW/cm^2	30 μW/cm^2	Continuous
Industrial	100 mW/cm^2	1–10 mW/cm^2	
Ambient RF energy	0.3 μW/cm^2	0.1 μW/cm^2 (cell phone)	Continuous

Source: Vullers, R.J.M., Schaijk, R.V., Doms, I., Van Hoof, C., and Mertens, R., *Solid-State Electron.*, 53, 684–693, 2009; Kim, S., Vyas, R., Bito, J., Niotaki, K., Collado, A., Georgiadis, A., and Tentzeris, M.M., *Proc. IEEE*, 102, 1649–1666, 2014.

Several EH techniques and their performances are listed in Table 1.1. It is observed here that the conversion efficiency in the ambient RFEH is highest compared to all other techniques, and it is a continuous source of energy.

In recent decades, the increase in applications of wireless technology has increased the utilization of various portable and handheld devices. Hence, to satisfy people's demand, a greater number of electromagnetic energy radiators, such as cellular systems, FM radio stations, wireless communication systems, digital television (DTV) towers, and Wi-Fi hubs, operating under different frequencies have been established. This increases the density of available power in the surrounding environment. Thus, RFEH technology has attracted the globe over all other EH techniques due to its increasing availability and power density.

The RFEH has the capability of supplying power to batteries and electronic devices through a wireless media; this is specifically convenient in isolated areas, especially where the accessibility is a difficulty or where substituting batteries is difficult (e.g. aircraft, chemical implants, bridges). Therefore, instead of using conventional batteries that have an inadequate lifetime and also cause pollution to the surrounding environment, RF energy harvesters are able to generate electrical power in a safe, cleanly, and cheap manner. A prominent benefit of RFEH is the capacity to renovate ambient energy into electrical power all over the day and night and in both indoor and outdoor environments. Although microwave signals in the ambient environment are very low in density as compared to solar and wind energies, the RF signals are available at all times. RFEH techniques allows to reduce the size of harvesting system (compared to wind turbines and PV cells in applications) if necessary. Hence, RFEH could be a better solution to deliver a sustainable energy source to meet upcoming difficulties.

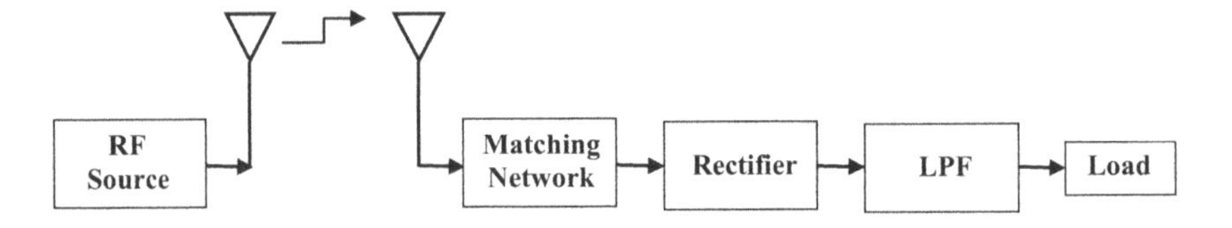

FIGURE 1.1 RFEH process by using a rectenna system. (From Brown, W.C., *J. Mirow. Power*, 1, 1–20, 2016. With permission.)

The general architecture of the rectifying antenna is shown in Figure 1.1. The rectenna captures RF signals from various sources in the ambient environment and converts them into electrical energy. The output from the rectenna system can be distributed to various applications such as implantable devices and radio frequency identification (RFID) systems. The rectenna basically comprises an antenna, rectifier circuit, low pass filter (LPF), and a matching network. The antenna acts as a transducer that receives RF signals from the surrounding environment, a rectifier circuit is used to convert the received ambient signals into electrical DC power, a matching network is used to transfer the maximum amount of received power by the antenna to the rectifier circuit, and an LPF circuit allows DC signals pass through the output load terminals of the rectenna system.

The rectenna was initially designed for wireless power transmission (WPT) applications. The wireless energy transmission concept was first introduced by Brown [28] in the 1960s by combining antenna and rectifier in order to receive the high-frequency electromagnetic energy beam (Figure 1.2). The proposed techniques of the rectenna were demonstrated in 1973 [29] for powering a helicopter through an electromagnetic beam from the earth (Figure 1.3). The rectenna circuit also finds its implementation in wireless energy harvesting (WEH) applications. WEH is the process in which the ambient wireless energy is extracted and converted to DC signals

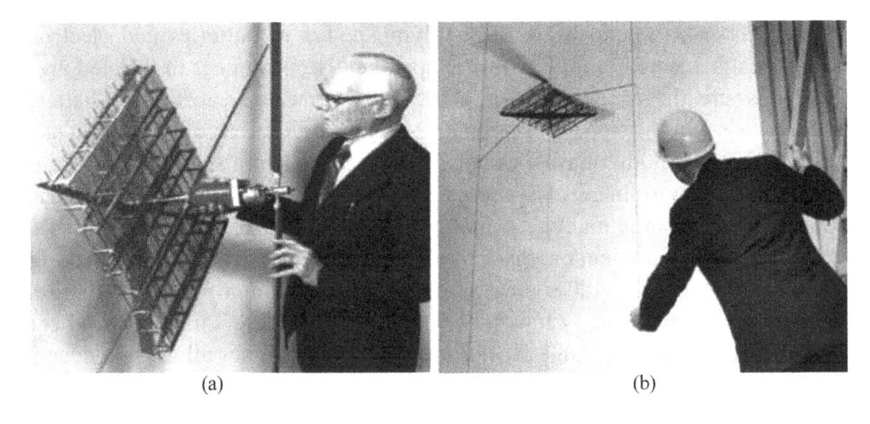

FIGURE 1.2 Microwave-powered helicopter system: (a) close view and (b) helicopter view. (From Brown, W.C., *J. Mirow. Power*, 1, 1–20, 2016. With permission.)

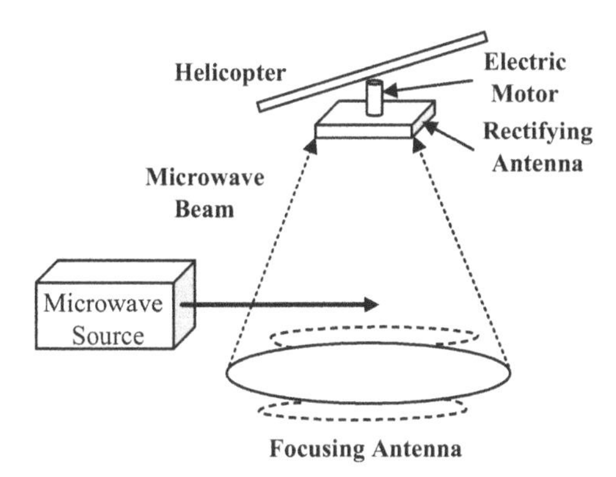

FIGURE 1.3 Basic design elements of microwave-powered helicopter. (From Brown, W.C., Mims, J.R., and Heenan, N.I., "An experimental microwave-powered helicopter," *1958 IRE International Convention Record*, 225–236, 1966. With permission.)

to operate low-power electronic devices. For this reason, rectennas are required in EH systems. However, designing the rectenna for EH applications poses many challenges. The rectenna is also used in solar power transmission (SPT) applications. The rectenna is placed on the earth over a larger area to collect the EM energy transmitted from the space. The energy collected by the rectenna is converted into DC, and again this DC power is converted into AC and then distributed through different transmission lines. The rectenna also find its applications in wireless sensor networks. The rectenna is integrated into a sensor element to operate efficiently without any power fluctuation problems. The rectenna harvests energy from the sources available in the surrounding environment or specially designed sources. A short-range wireless power transmission and reception system and method, as discussed in Ref. [30], is shown in Figure 1.4, and the power requirements of some portable devises are shown in Figure 1.5 [31].

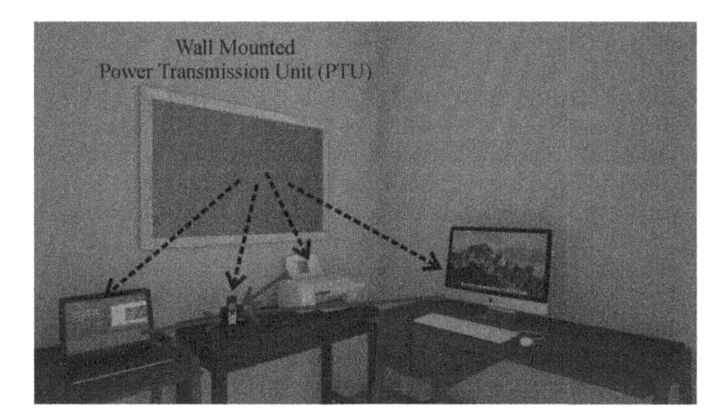

FIGURE 1.4 Wall-mounted power transmission units (PTUs). (From Vecchione, E. and Keegan, C., "Short-range wireless power transmission and reception," US 2006/0238365A1, 1–10, 2006. With permission.)

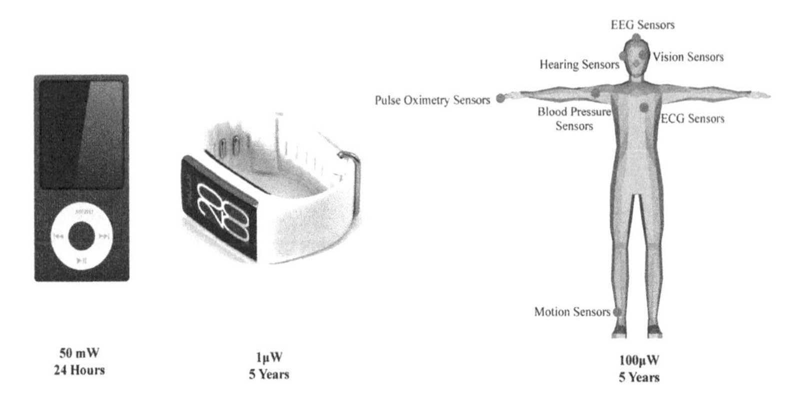

FIGURE 1.5 Power requirements of some portable devices. (From Vullers, R.J.M., Schaijk, R.V., Visser, H.J., Penders, J., and Hoof, C.V., *IEEE Solid-State Circuits Magazine*, 2, 29–38, 2010. With permission.)

REFERENCES

1. R. Singh, N. Gupta, and K.F. Poole, "Global green energy conversion revolution in 21st century through solid state devices," *Proceedings of the 26th International Conference on Microelectronics* (MIEL 2008), Nis, Serbia & Montenegro, pp. 45–54, May 2008.
2. S. Hemour and W. Ke, "Radio-frequency rectifier for electromagnetic energy harvesting: development path and future outlook," *Proceedings of the IEEE*, vol. 102, pp. 1667–1691, November 2014.
3. A. Varpula, S.J. Laakso, T. Havia, J. Kyynarainen, and M. Prunnila, "Harvesting vibrational energy using material work functions," *Scientific Reports*, vol. 4, no. 6799, pp. 6799, 2014.
4. A. Khaligh, P. Zeng, and C. Zheng, "Kinetic energy harvesting using piezoelectric and electromagnetic technologies-state of the art," *IEEE Transactions on Industrial Electronics*, vol. 57, pp. 850–860, 2010.
5. Q. Zhang and E.S. Kim, "Vibration energy harvesting based on magnet and coil arrays for watt-level handheld power source," *Proceedings of the IEEE*, vol. 102, no. 12, pp. 1747–1761, 2014.
6. A.H. Slocum, G.E. Fennell, G. Dundar, B.G. Hodder, J.D.C. Meredith, and M.A. Sager, "Ocean renewable energy storage (ORES) system: analysis of an undersea energy storage concept," *Proceedings of the IEEE*, vol. 101, no. 4, pp. 906–924, April 2013.
7. J.A. Paradiso and T. Starner, "Energy scavenging for mobile and wireless electronics," *IEEE Pervasive Computing*, vol. 4, no. 1, pp. 18–27, January 2005.
8. R.J.M. Vullers, R.V. Schaijk, I. Doms, C. Van Hoof, and R. Mertens, "Micropower energy harvesting," *Solid-State Electronics*, vol. 53, pp. 684–693, 2009.
9. P. Jaffe and J. McSpadden, "Energy conversion and transmission modules for space solar power," *Proceedings of the IEEE*, vol. 101, pp. 1424–1437, 2013.
10. V. Raghunathan, A. Kansal, J. Hsu, J. Friedman, and M. Srivastava, "Design considerations for solar energy harvesting wireless embedded systems," *IPSN'05 Proceedings of the 4th International Symposium on Information Processing in Sensor Networks*, Boise, ID, USA, USA, Article No. 64, pp. 457–462, 2005.
11. M.A. Green, K. Emery, Y. Hishikawa, W. Warta, and E.D. Dunlop, "Solar cell efficiency tables (Version 38)," *Progress in Photovoltaics: Research and Applications*, vol. 19, pp. 565–572, 2011.

12. S.H. Krishnan, D. Ezhilarasi, G. Uma, and M. Umapathy, "Pyroelectric-based solar and wind energy harvesting system," *Sustainable Energy, IEEE Transactions on Sustainable Energy*, vol. 5, pp. 73–81, 2014.
13. S. Kim, R. Vyas, J. Bito, K. Niotaki, A. Collado, A. Georgiadis, and M.M. Tentzeris, "Ambient RF energy-harvesting technologies for self-sustainable standalone wireless sensor platforms," *Proceedings of the IEEE*, vol. 102, pp. 1649–1666, 2014.
14. L. Long and H. Ye, "How to be smart and energy efficient: a general discussion on thermochromic windows," *Scientific Reports*, vol. 4, no. 6427, pp. 1–10, 2014.
15. H. Lhermet, C. Condemine, M. Plissonnier, R. Salot, P. Audebert, and M. Rosset, "Efficient power management circuit: from thermal energy harvesting to above-IC microbattery energy storage, " *IEEE Journal of Solid-State Circuits*, vol. 43, no. 1, pp. 246–255, 2008.
16. Y. Du, K. Cai, S. Chen, H. Wang, S.Z. Shen, R. Donelson, and T. Lin, "Thermoelectric fabrics: toward power generating clothing," *Scientific Reports*, vol. 5, pp. 6411, 2015.
17. S.B. Inayat, K.R. Rader, and M.M. Hussain, "Nano-materials enabled thermoelectricity from window glasses," *Scientific Reports*, vol. 2, no. 841, pp. 1–7, 2012.
18. H.S. Kim, J.H. Kim, and J. Kim, "A review of piezoelectric energy harvesting based on vibration," *International Journal of Precision Engineering and Manufacturing*, vol. 12, no. 6, pp. 1129–1141, December 2011.
19. T.L. Chern, P.L. Pan, Y.L. Chern, W.T. Chern, W.M. Lin, C.C. Cheng, J.H. Chou, and L.C. Chen, "Excitation synchronous wind power generators with maximum power tracking scheme," *IEEE Transactions on Sustainable Energy*, vol. 5, no. 4, pp. 1090–1098, 2014.
20. F. Kong, C. Dong, X. Liu, and H. Zeng, "Quantity versus quality: optimal harvesting wind power for the smart grid," *Proceedings of the IEEE*, vol. 102, no. 11, pp. 1762–1776, 2014.
21. D. Jang, J. Jeon, and S.K. Chung, "Acoustic energy harvester utilizing a miniature rotor actuated by acoustically oscillating bubbles-induced synthetic jets," *2017 IEEE 30th International Conference on Micro Electro Mechanical Systems* (MEMS), Las Vegas, NV, USA, pp. 41–44, January 2017.
22. H. Basaeri, Y. Yu, D. Young, and S. Roundy, "A MEMS-scale ultrasonic power receiver for biomedical implants," *IEEE Sensors Letters*, vol. 3, no. 4, pp. 1–4, April 2019.
23. B. Li, and J.H. You, "Harvesting ambient acoustic energy using acoustic resonators," *Proceedings of Meetings on Acoustics*, vol. 12, pp. 1–8, 2011.
24. A. Costanzo, M. Dionigi, D. Masotti, M. Mongiardo, G. Monti, L. Tarricone, and R. Sorrentino, "Electromagnetic energy harvesting and wireless power transmission: a unified approach," *Proceedings of the IEEE*, vol. 102, no. 11, pp. 1692–1711, 2014.
25. W.C. Brown, "The history of power transmission by radio waves," *IEEE Transactions on Microwave Theory and Techniques*, vol. MTT-32, no. 9, pp. 1230–1242, 1984.
26. W.C. Brown, "Experiments involving a microwave beam to power and position a helicopter," *IEEE Transactions on Aerospace and Electronic Systems*, vol. AES-5, no. 5, pp. 692–703, 1969.
27. N. Shinohara, "Power without wires," *IEEE Microwave Magazine*, vol. 12, pp. S64–S73, 2011.
28. W.C. Brown, "The microwave powered helicopter," *Journal of Microwave Power*, vol. 1, pp. 1–20, 2016.
29. W.C. Brown, J.R. Mims, and N.I. Heenan, "An experimental microwave-powered helicopter," *1958 IRE International Convention Record*, New York, NY, USA, USA, pp. 225–236, March 1966.
30. E. Vecchione and C. Keegan, "Short-range wireless power transmission and reception," US 2006/0238365A1, pp. 1–10, October 2006.
31. R.J.M. Vullers, R.V. Schaijk, H.J. Visser, J. Penders, and C.V. Hoof, "Energy harvesting for autonomous wireless sensor networks," *IEEE Solid-State Circuits Magazine*, vol. 2, pp. 29–38, 2010.

2 Antennas for RFEH Systems

2.1 INTRODUCTION

As receiving antenna in the RFEH is very important component to capture the RF energy from surrounding environments, the antenna innovation works involve investigations of new antenna variations to capture maximum RF energy, such as circularly polarized (CP) antennas; high-gain, wideband, array antennas; all/dual-polarized antennas; small low-power antennas; efficient antennas; narrow band/multiband antennas; and opaque or transparent antennas. Antennas with different shapes and types have been employed in RFEH applications, from the simple monopole/dipole to more complex designs such as the 3D or array or bow-tie or spiral antenna. This chapter presents extensive literature review on current state-of-the-arts antennas for RFEH systems.

2.2 CIRCULARLY POLARIZED ANTENNAS

The circularly polarized (CP) antennas allow the system to harvest RF energy regardless of the device orientation as well as different polarized RF waves [1]. As the RF energy found in the surrounding environment can exist in any orientation/polarization and phase alignment or unknown RF sources, CP antennas are more desirable for RFEH systems. Capability to radiate and receive RF energy in any plane with minimum loss makes CP antennas a key candidate for RFEH system.

It improves the power conversion efficiency (PCE) of the RFEH system by reducing polarization mismatch losses. Corner truncation, spur lines, slits, stub loading, and cross-shaped slots are the few methods used to achieve CP radiation. For single-feed technique [2,3], the antenna is fed at 45° with respect to the perturbation for achieving CP radiation. These types of antennas are simple and compact in structure, but have narrow bandwidth. A dual-feed structure provides a wider CP (3-dB axial ratio) bandwidth compared to the single-feed antenna, but it involves a bulky feeding system with ground plane. For single-feed CP antenna, the slight perturbing of the radiator structure is required to excite two approximate same amplitude modes with phase shift of 90°. For square patch radiator, different perturbation methods such as truncated corners, slit/slits, slot/slots, and stub/stubs are employed to generate CP radiation. It leads to asymmetric current path along the diagonal length that produces two different resonance frequencies at 90° phase difference generating CP [2] but at the cost of reduced bandwidth and efficiency. Nasimuddin and co-researchers have developed different types of slit/slotted CP antennas for RFEH systems [1,4–10].

Stacked structures, artificial magnetic conductors (AMCs), and metamaterials are commonly used methods to overcome the narrow bandwidth limitation and gain enhancement [11,12]. However, one of the widely used methods to improve the bandwidth of CP antenna is the stacked patch concept. In this configuration, the parasitic patch element over the patch radiator is used [12]. A wide bandwidth is achieved through the electromagnetic (EM) coupling between the driven (main radiator) and the parasitic elements for generating resonances that are close to each other, but the limitation is the increased overall size.

The annular-slot ring around a circular patch for CP is reported in Ref. [13], which achieves 25% RF-to-DC conversion efficiency at 0-dBm input power but reduces drastically at –20 dBm input power. It uses thin and flexible substrate and suffers from bending losses, which reduces the efficiency.

Wide-beam CP radiation enhances the coverage area of the RFEH systems. The CP radiation coverage (3-dB AR beamwidth) can be enhanced using different techniques [1,6] such as tapered-slit patch radiator and integrated stubs/small-circular-patches patch radiator antennas.

The omnidirectional CP antenna can be achieved using the truncated corner of square patch radiator topology and multilayer structure antenna [14], as its wide angle CP coverage can receive RF waves from all directions with different polarized waves. The antenna has bandwidth of 3.6%, 2.41 GHz (2.43–2.45 GHz), and when the antenna is used as a receiver in RFEH system, the output voltage at frequency of 2.40 GHz is 255 mV. In Ref. [8], an E-shaped slot is investigated along the orthogonal axis of the circular patch to generate CP performance, and the antenna has wide-angle CP radiation of 140° with a bandwidth of 3.2% and a gain of more than 5.0 dBic at 2.38 GHz. Four stubs are integrated at four corners of a square patch antenna to achieve a wide-beam CP coverage in Ref. [6]. A maximum 10-dB bandwidth of 4.8% at 2.5 GHz is obtained for the asymmetric gap case. The corresponding maximum gain is 4.5 dBic, and the average output voltage of 1.5 mV is achieved at different rotation angles. A wideband and high-gain CP antenna is reported in Ref. [9], and CP is generated by cutting a Teo-shaped slot at the square patch radiator, while aperture coupling is used to broaden the bandwidth. A 10-dB bandwidth of 33% (1.89–2.66 GHz) with 3-dB axial ratio bandwidth of 100 MHz (2.4–2.5 GHz) and maximum gain of 6.8 dBic are achieved at 2.28 and 2.5 GHz, respectively. The antenna has a good performance in terms of size, bandwidth, and gain. However, the aperture coupling is a disadvantage from fabrication point of view. Rahim et al. [15] have presented a new design of harmonic suppression microstrip antenna with the CP properties at 900 MHz for an ambient RFEH system by truncating the corner of the rectangular microstrip antenna for achieving the required axial ratio below 3 dB. A compact dual-fed CP antenna array with reflector is designed by Ismail et al. [16] for RFEH system comprising four patches with dual slots placed underneath of radiating patches to reduce the antenna's sizes. A dual-polarized, dual-frequency, and dual-access antenna for RF energy recycling applications is presented by Haboubi et al. [17]. The antenna shows good measured performance at 940 MHz and 2.45 GHz with a simulated broadside gain higher than 6 dBi and a return loss lower than –12 dB. It also exhibits harmonic rejection properties at 1.88 and 4.9 GHz avoiding the use of RF filters when the antenna is connected to a non-linear device as the diode rectifier because no RF filter is required to suppress higher order harmonics. Ghosh et al. [18] have presented a design of multiband microstrip monopole antenna with CP and impedance of 377 Ω to grab the ambient RF energy efficiently using the electromagnetic radiation from cell towers in the CDMA and GSM bands. Ghosh [19] has presented the design of an efficient and simple RFEH rectenna module consisting of a slotted-antenna with CP and bridge rectifier operating in GSM 900 band for energizing the wireless sensor networks (WSN). Jie et al. [10] have proposed a proximity-coupled (electromagnetic coupling feed) CP slotted-circular microstrip antenna for RFEH applications where two asymmetric-circular slots are cut diagonally onto a circular-patch antenna to generate two approximate equal amplitude modes with a 90° phase-shift for requirements of CP radiation. A six-band frequency independent dual-CP planar log periodic antenna, with high gain and the best CP coverage (axial ratio below 3 dB), for ambient RFEH was presented by Khaliq et al. [20] to harvest energy from six bands within the frequency range from 650 MHz to 2.5 GHz that covers various wireless standards (digital TV band, UHF-band, cellular bands, and Wi-Fi bands). Adam et al. [21] have proposed a novel wideband CP antenna with high gain that is suitable to harvest energy in which the wideband CP is achieved by implementing EM-coupled feed with a defected ground plane structure. Some CP antenna designs for RFEM systems such as compact, high gain, wideband, high conversion efficiency, widebeam CP coverage, and enhanced AC-to-DC conversion efficiency are compared in Table 2.1.

A novel six-band dual-CP antenna [27] has been proposed to receive the RF energy from six different frequency bands with almost all polarized waves, as shown in

TABLE 2.1
Comparison of CP Antenna Designs for RFEH Systems

Antenna Structure	Frequency	Gain (dBic)	Bandwidth	Remark
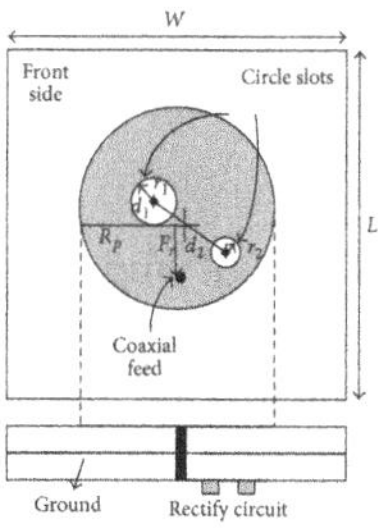 [22]	2.45 GHz	3.36	137 MHz	Circular slotted-patch antenna, efficiency degradation due to antenna miniaturization
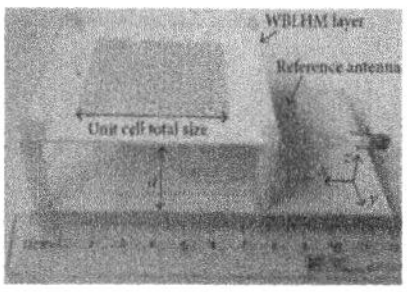 [23]	6.0 GHz	12.0	38.7%	Array antenna with metasurface used to improve gain and bandwidth

(*Continued*)

TABLE 2.1 (*Continued*)
Comparison of CP Antenna Designs for RFEH Systems

Antenna Structure	Frequency	Gain (dBic)	Bandwidth	Remark
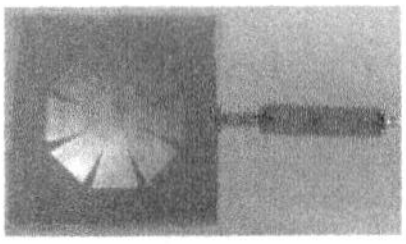 [1]	0.9 GHz	5.8	7.0 MHz	Slit-patch antenna with metamaterial based rectifier and 3-dB AR widebeam of 180°
 [24]	2.5 GHz	4.5–6	400 MHz	Widebeam of 179° to impedance matching and improved front-to-back ratio
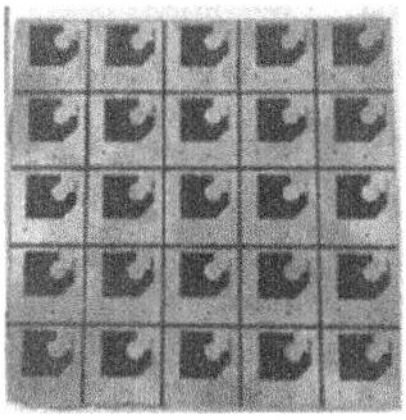 [25]	5.8 GHz	6.9	31.8%	Maximum conversion efficiency of 62% with harmonic suppression
 [26]	0.915 GHz	-	-	AC-to-DC conversion efficiency of 82%

Figure 2.1. Due to the nonlinearity and complex input impedance of the rectifying circuit, the design of a multiband and/or broadband rectenna is always challenging and its performance can be easily affected by variation in the input power level and load. A broadband dual-CP receiving antenna, which has a very wide bandwidth (from 550 to 2.5 GHz) and a compact size, was developed. Millimeter-wave CP rectenna [28] has been developed at 24 GHz to harvest energy from millimeter-wave sources.

2.3 HIGH-GAIN ANTENNAS

Key goal of antenna technology for RFEH system is to achieve high gain with compact size. An antenna is a key component for capturing RF signal. In RFEH systems, a high-gain receiving antenna can enhance the received power but at the cost of its required large antenna aperture or bulky antenna structure. Various types of high-gain

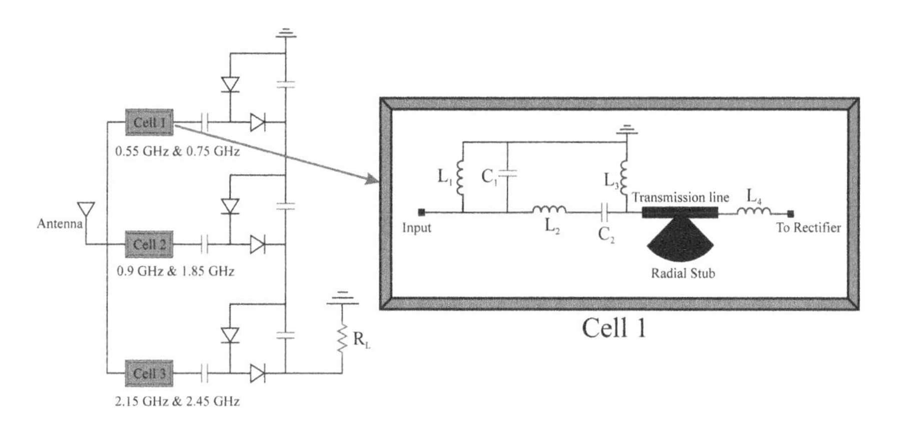

FIGURE 2.1 Six-band dual-CP antenna for RFEH system. (From Song, C., Huang, Y., Carter, P., Zhou, J., Yuan, S., Xu, Q., and Kod, M., *IEEE Trans. Antenn. Propag.*, 64, 3160–3171, 2016. With permission.)

antennas have been used for RFEH applications in the literature. The suitability of using a printed circuit board (PCB) patch receiving antenna for RFEH to power a wireless soil sensor network deployed in an outdoor environment has been investigated by Sim et al. [29] based on a chosen PCB material as the antenna substrate by presenting an enhanced gain circular patch with a ring-shaped parasitic radiator. Marshall and Durgin [30] have introduced the theory of N-by-N staggered pattern charge collectors (SPCC) and a methodology to design and optimize SPCCs for maximum efficiency of the RFEH system which uses multiple sub-arrays to form an aggregate gain pattern for harvesting RF energy more efficiently than a single antenna or a collection of antennas occupying a similar footprint when the transmitter location is unknown.

Khayari et al. [31] have proposed a design capable of charging small power consumption sensors comprising high-profile monopole antenna array with a reflector by capturing ambient RF waves for enhancing efficiency of the energy harvesting devices. Kang et al. [32] have proposed a RFEH technique using planar circular spiral inductor antenna in which a single element circular spiral antenna and a 2×1 array is designed to determine the feasibility to harvest power density produced by the proposed antenna. Shen et al. [33] have presented the application of metamaterial-loaded Vivaldi antenna in RF energy harvester in which the Vivaldi antenna is loaded with anisotropic zero-index metamaterial (ZIM), designed to enhance the radiation gain.

For comparison of the size, −10-dB impedance bandwidth, gain, and radiation patterns, a 2×2 array of spiral resonators and a 2×2 array electromagnetic band gap (EBG) structure using aperture coupled have been proposed by Olule et al. in Ref. [34]. The design of miniature high-gain dielectric resonator antenna (DRA) for RFEH application with high figure of merit to increase the power received using numerical approximation to assist the antenna design and modelling has been proposed in Ref. [35] by Masius and co-workers. For wideband with high gain, a hybrid DRA [36] has been investigated for RFEH application, as shown in Figure 2.2. The antenna consists of a simple rectangular DRA backed by a rectangular slot in the ground plane. An inverted T-shaped feed line is used to excite DRA and rectangular

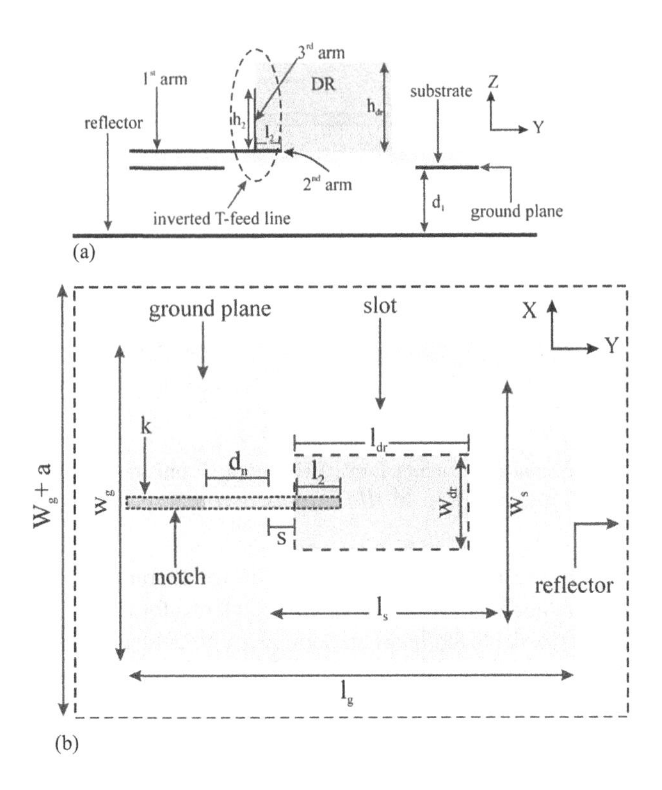

FIGURE 2.2 A wideband high-gain hybrid antenna (a) cross-sectional view, (b) bottom view. (From Agrawal, S., Gupta, R.D., Parihar, M.S., and Kondekar, P.N., *Int. J. Electron. Commun.*, 78, 24–31, 2017. With permission.)

slot simultaneously to enhance the bandwidth, and a metallic reflector is placed behind the antenna to enhance the antenna gain. Measured results show that the proposed antenna offers an impedance bandwidth of about 120% (1.67–6.7 GHz) with maximum gain of 9.9 dBi and can achieve the rectenna efficiency of 61%.

Recently, the high-gain grid array antenna (GAA) based rectennas [37,38] have been proposed for RFEH in low power density environment, as shown in Figure 2.3. The GAA without any extra beamforming networks/without phase shifters can realize beam steering with high gain to capture more RF waves. Benefited from the capability of tilting the beam angle, two isolated ports at two opposite edges of GAA can excite two beams.

2.4 WIDEBAND ANTENNAS

Wideband receiving antennas are very useful to collect the RF energy from wide spectrum (wide frequency range) in the surrounding environment. Different types of wideband antennas are investigated for RFEH system. Devi et al. [39] have presented and discussed the design of a 377-Ω E-shaped patch antenna with partial ground plane, in which two parallel slots are introduced into the conventional patch antenna, and it

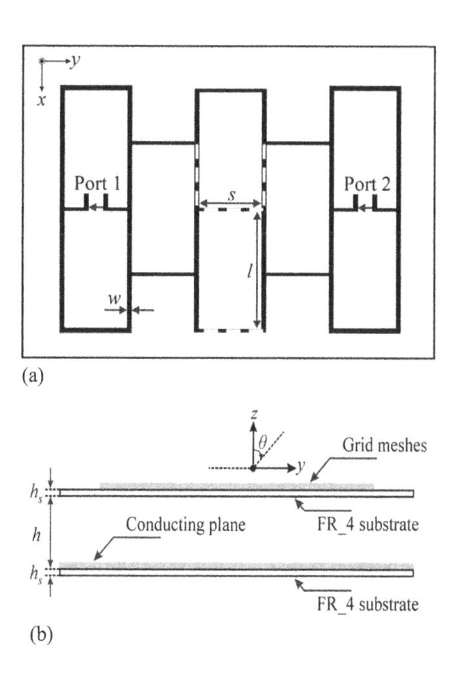

FIGURE 2.3 A high-gain GAA-based rectenna for RFEH system (a) top-view (b) side-view. (From Hu, Y.Y., Sun, S., Xu, H., and Sun, H., *IEEE Trans. Micro. Theory Tech.*, 67, 402–413, 2019. With permission.)

is investigated by the currents through the patch to enhance its bandwidth. Zainuddin et al. [40] have presented a wideband antenna of Ice-cream Cone structure, and it is integrated with a rectifying circuit which has the potential to be used for RFEH system and fabricated on a double-sided FR4 substrate using an etching technique. Peter et al. [41] have proposed a novel transparent ultra-wideband (UWB) antenna for photovoltaic solar-panel integration and RFEH, in which a transparent cone-top-tapered slot antenna covering the frequency range from 2.2 to 12.1 GHz is designed and fabricated to provide UWB communications while integrated into solar panels as well as harvest EM waves from free space and convert them into electrical energy.

Rai et al. [42] have presented an E-patch antenna with improved bandwidth and efficiency using a technique to make trade-offs required for widening the frequency band without variation in efficiency. Saghlatoon et al. [43] have presented a fully inkjet-printed novel wideband planar monopole antenna on thin packaging cardboard available in bulk for low-cost and environmentally friendly mass manufacturing using thin dielectric coating on the cardboard. Mrnka et al. [44] have presented a wideband DRA based on identical bow-tie DRs with different feeding schemes for RFEH covering three main wireless communication systems, namely global system for mobile communications (GSM) operating on 1800 MHz, universal mobile telecommunications system (UMTS) at 2100 MHz, and wireless local area network (WLAN) at 2400 MHz. Wen et al. [45] have presented a wideband collar-shaped antenna consisting of two folded dipoles, which are laid on a circular FR-4 substrate, designed for multiband RFEH system. Iwata et al. [46] have designed a wideband and high-gain antenna integrated with a multilayer insulation blanket and fabricated by inkjet printing technique for RFEH system.

A novel Cantor fractal antenna has been designed by Bakytbekov et al. [47] to obtain decent gain performance and an omnidirectional radiation pattern, realized through a combination of 3D inkjet printing of plastic substrate and 2D inkjet printing of metallic nanoparticles-based ink. Agrawal et al. [36] have designed and experimentally validated a novel high-gain and broadband hybrid DRA by combining the DRA and an underlying slot with a narrow rectangular notch which effectively broadens the impedance bandwidth by merging the resonances of the slot and DRA. Lee et al. [48] have presented the development of a hybrid antenna for electromagnetic energy harvesting and microwave power transmission (MPT), designed by mounting a 2×2 array resonant loop antenna on a wideband tapered monopole antenna with an etched hole and a patch on the other side. Vincetti et al. [49] have presented a broadband UHF antenna with high inductive input impedance for RFEH consisting of a small feeding loop and a biconical radiating dipole fabricated on a FR4 substrate. Wang et al. [50] have designed and tested a broadband RFEH system across a frequency range of 900 MHz–3 GHz, in which a planar monopole antenna fed by coplanar waveguide (CPW) is used. Bertacchini et al. [51] have presented a RFEH comprising a differential radio frequency-to-direct current (RF-DC) converter realized in ST130 nm complementary metal-oxide-semiconductor (CMOS) technology and a customized broadband PCB antenna with inductive coupling feeding.

The performance of a broadband log periodic toothed trapezoidal antenna designed to work from 2.4 to 5.2 GHz but characterized from 400 MHz to 2.4 GHz was proposed by Simon et al. [52] to know its potential for RFEH system. Arrawatia et al. [53] have presented a broadband bent triangular omnidirectional antenna for RFEH to receive both horizontal and vertical polarized waves, and it has a stable radiation pattern across the entire bandwidth. Maher et al. [54] have proposed a broadband planar antenna impedance matched with a 50 Ω line across the frequency band from 2.1 to 7 GHz for RFEH, in which three deep resonances have been introduced at the frequencies 2.4, 4, and 5.8 GHz to increase the utility of the power resulted from the operating applications at these frequencies. Rivera et al. [55] have presented experimental results of an antenna array for RFEH, in which an array of five high monopoles was implemented and its RF receiving properties were measured. Bai et al. [56] have proposed a novel broadband CPW-fed fractal antenna for RFEH, in which the equilateral triangle slot is made to form a 2-iteration fractal antenna based on circular patch.

2.5 MULTIBAND ANTENNAS

To improve efficiency of the RFEH system, the receiving antenna in the RFEH system should capture as much as RF energy with different RF sources, so the multiband receiving antenna is a very useful candidate. Different types of the multi-band antennas such as dual-band, tri-band, and multi-frequency have been developed for RFEH applications. Dual-band antenna based RFEH system can capture the RF power from two frequency band sources to enhance the harvesting power. Different types of dual-band antennas have been designed for RFEH systems.

Shaoa et al. [57] have presented a PCB-based planar dual-band monopole antenna design that can provide the required impedance matching to the RF chip at two bands and can be easily integrated with the RF harvester chip on the same circuit board. Li et al. [58] have presented the development of a dual-band energy harvester that can achieve optimum PCE at both 1.8 and 2.4 GHz, in which the antenna and voltage multiplier impedance are designed to be complex conjugate to each other for maximum power transfer as well as voltage boosting. Arrawatia et al. [59] have presented the RFEH system extracting energy from RF radiation for battery charging applications comprising a new differential center tapped microstrip antenna, off-chip matching circuit, on-chip novel CMOS rectifier, and control circuitry in 180 nm CMOS technology. Saraiva et al. [60] have experimentally characterized the developed wearable antennas and circuits for RFEH system, which consist of a 5-stage Dickson voltage multiplier with an impedance matching circuit, rectifier, and energy storage sub-system. Zakaria et al. [61] have investigated the dual-band monopole antenna based on meander line structure using CST software for GSM band applications for RFEH systems. Barreca et al. [62] have identified the spectrum opportunities for RFEH through power density measurements from 350 MHz to 3 GHz and proposed a dual-band band printed antenna operating at GSM bands (900 MHz/1800 MHz) with gains of the order 1.8–2.06 dBi and efficiency 77.6%/84.0%.

Barcak and Partal [63] have designed and implemented a matched multiband microstrip antenna intended to collect RF energy efficiently in commonly used portions of the selected cellular frequency spectrum including the ISM bands. Li et al. [64] have proposed an antenna and harvester co-design methodology to improve RF-to-DC energy conversion efficiency using an IBM 0.13 m process obtained over 9% efficiency for two different bands (around 900 MHz and around 1900 MHz) at an input power as low as 19.3 dBm. Schematic model of RFEH circuit coupling with a multiple ISM band microstrip antenna has been presented by Alam and Moury [65] to empower wireless sensors operating at 2.4 and 5.8 GHz by tuning its L-C impedance matching network. Borges et al. [66] have presented RFEH circuits specifically developed for GSM bands (900/1800) and a wearable dual-band antenna suitable for possible implementation within clothes for body-worn applications where the PCB fabrication is achieved by means of a rigorous control in the photo-positive method and chemical bath procedure is applied to the PCB for attaining better values for the conversion efficiency. Salih and Sharawi [67] have designed a dual-band electrically small meander line antenna for RFEH applications, in which the dual-band property is achieved by using two different radiators, each of them is responsible for covering one band and a $\lambda/4$ transformer is used to match the antenna at 915 MHz. Borges et al. [68] have showcased the most recent advances toward the development of efficient RFEH systems to power wearable biomedical nodes, in which different and important observations have been reported concerning the different RFEH devices as well as the energy storing systems that can be used to store the energy for posterior use by a WBAN node. Kamoda et al. [69] have studied loop antennas over AMC surfaces with the objective of designing a dual-band RFEH antenna and implemented AMC surface and the loop antenna as a dual-band RFEH panel together with RF-to-DC conversion circuits and a power management circuit. Arrawatia et al. [70] have designed, fabricated, and tested a differential microstrip antenna with a gain of 8.5 dBi and a bandwidth of 135MHz, in which the antenna can be used either in center grounded or in differential configuration. Zhao et al. [71] have presented a simple dual-band patch antenna that radiates at 920MHz and 2.4GHz effectively, designed with meander slot and shorting probe to increase the current path and reduce the resonant frequency. Zhou et al. [72] have designed a multi-band fractal antenna for RFEH based on a general cross-dipole antenna where iterated slot is added on each pole to achieve a wide multi-band performance and a derivative version of the antenna is developed by surrounding the original antenna with a ring structure for a better 900MHz performance. Pal and Choukiker [73] have presented the RFEH with frequency reconfigurable microstrip patch antenna for two different frequency bands, reconfigurable at 1.6 and 2.4GHz using RF PIN diodes. Hebelka et al. [74] have analyzed a new type of a Koch-like antenna designed for RFEH specifically covering two different frequency bands (GSM 900 and Wi-Fi). Mouris et al. [75] have studied a dual-band antenna and a rectifying circuit that can be integrated together for the operation of an implantable glucose sensor. Rectenna consists of a capacitively-loaded printed inverted F-antenna (IFA) on a flexible thin Liquid Crystal Polymer Rogers substrate with a Schottky-diode based single stage rectifying circuit and it can work dual-band (1.6 GHz and 2.4 GHz) with high RF-to-DC conversion efficiency. Mathur et al. [76] have presented the new structure of coplanar monopole receiving antenna with array arrangement of 2×2 coplanar monopole antenna for the RFEH purposes, designed to work on almost all useful multi frequency bands (i.e., Radio, GSM, ISM, and UWB). Sarma and Akhtar [77] have proposed a high-gain printed dual-band meandered dipole antenna for RF energy harvesting applications in the GSM-1800 and WiFi (2.45GHz) bands utilizing a meandered structure in order to achieve dual-band properties as well as size miniaturization. Xu et al. [78] have proposed a new 2×2 circular microstrip antenna array with air dielectric layer for ambient RFEH , in which two pairs of arc-shaped slots located close to the boundary of the circular microstrip patch have been designed for making dual band and extending the frequency bandwidth. Raj [79] has presented a MEMS Based multi-band RFEH scheme by operating wireless sensor networks along with ultralow power electronic devices for various applications accomplished by having antennas, rectifier circuits for conversion of RF signals into DC, super capacitor, boost converters, and necessary power management circuits. Bakkali et al. [80] have focused on ambient RF energy available from commercial broadcasting stations in order to provide a system based on RFEH using a new design of receiving antenna, aimed at greatly increasing the RFEH efficiency across the Wi-Fi bands: 2.45 and 5GHz. Loss et al. [81] have presented a smart coat with a dual-band textile antenna for RFEH, operating at GSM 900 and DSC 1800 bands, which is fully embedded in the garment. Azam et al. [82] have designed Dickson charge pump circuit paired with a compact, multi-resonant loop antenna using surface mount components for RFEH at 900MHz and 2.4GHz. A multi-band planar antenna consisting of a monopole with elliptical arc on the top metal layer and modified ground plane on the other side is proposed and designed by Elsheakh [83] for RFEH applications. Hafeez and Jilani [84] have proposed a novel flexible antenna consisting of a triangular patch printed on a flexible PET substrate for RFEH applications, in which defected ground structures (DGS) have been introduced to generate multiple resonances for improving bandwidth.

Yunus et al. [85] have presented microstrip antenna design of a dual-band RFEH system for self-powered devices, designed using a slot-coupled structure to realize for dual microwave bands operation by using an optimization parameter of 50 Ω feed point on microstrip line. Arun and Marx [86] have designed a novel micro-controller-based frequency reconfigurable patch antenna using RF PIN diodes by adjusting the ON and OFF state of the diodes using the microcontroller unit.

Agrawal et al. [87] have proposed a compact planar CPW-fed antenna including five radiating elements and a stepped ground plane operating across the four frequency bands, GSM-900, GSM-1800, Wi-Fi 2.1 GHz, and Bluetooth (2.4 GHz), for RFEH applications. Patil and Gahankari [88] have designed a dual-band microstrip antenna suspended above ground plane to operate in GSM 1.8 and ISM 2.4 GHz bands, in which the antenna incorporates the capacitive feed strip which is fed by coaxial probe technique where the inductive impedance of probe is effectively cancelled by the capacitive patch. A new characterization and selection policy (CSP) using parallel and aggregated multi-armed bandit (PA-MAB) framework has been proposed by Darak [89] to aid optimum RF source selection in multi-antenna RFEH circuits. Tawk and Costantine [90] have proposed the design and dual functionality of two antenna arrays that operate from 1.8 to 2.4 GHz for either communication or RFEH where each antenna array is composed of two elements and integrated vertically on top of a solar panel. Kurvey and Kunte [91] have designed a planer stepped rectangular monopole antenna 1800 MHz for RFEH, in which initially, a rectangular monopole antenna resonating at GSM 1800, 3G, 4G, ISM band is designed, then steps are introduced in the planer monopole antenna near the feed line and the top side of rectangular monopole due to which the antenna starts resonating at GSM 900 along with present bands. Mathur et al. [92] have designed an efficient RFEH system using 21 antenna array structure of the coplanar monopole antenna which receives the RF and Microwave frequency signals present in the atmosphere and converts it into DC signal so that it can store in capacitor to eliminate the problem of continuous charging of a battery-operated electronics device. Celik and Kurt [93] have designed and simulated a multi-band fractal antenna for RFEH systems where FR4 glass epoxy material (1.6 mm thickness) is used for antenna design and coaxial feed connection is made from the center triangle of the patch. Muncuk and Alemdar [94] have presented a study of ambient RF signal strength distribution conducted in Boston city, indicating locations and associated RF bands that can point toward the practicality of ambient, RFEH and also demonstrated an adjustable circuit for harvesting from LTE 700 MHz, GSM 850 MHz, ISM 900 MHz bands with one single circuit.

2.6 HIGH-EFFICIENCY ANTENNAS

A high-efficient broadband rectenna [95] has been proposed for ambient wireless energy harvesting which has a novel broadband rectifying circuit with a new impedance matching circuit designed to match with the ambient RF signals with a relatively low power density. The power sensitivity has been improved by using a full-wave rectifier circuit configuration. A broadband dual-polarization cross dipole antenna has been designed to enhance the receiving capability of antenna. The harmonic rejection property has been embedded in an integrated antenna by using a novel slot-cutting approach in order to improve the overall efficiency and keep the overall size as small as possible. The simulated and measured results have shown that the rectenna has maximum conversion efficiency of around 55% for –10-dBm input power from 1.8 to 2.5 GHz. The power sensitivity is down to –35 dBm. The rectified DC power can be well above the incident power from any single resource due to the broadband operation and high efficient design. Considering the high DC power output of this design in a relatively low power density environment, this rectenna can be used for efficient wireless energy harvesting for a range of wireless sensor and network applications.

Mi et al. [96] have proposed and experimentally studied multiple RFEH antennas in essentially one space or area as a means to increase the energy or power/area ratio where four cooperating antennas are proposed in a square area that is less than twice the area required for a single antenna. Xie et al. [97] have presented a square antenna array for TV broadcast RFEH consisting of four square electrically small Egyptian-axe dipole (EAD) antennas, and each element comprises parasitic dipoles and exciting dipoles both with the right-angle arrow-shaped loadings, separately on two square-shaped substrates. Yang et al. [98] have proposed a 915-MHz RF-power emitter with an antenna array controlled by phase shifters to make directional energy beam point to the RFEH receiver. RF-power transmitter was designed with a 2×2 patch antenna array and tunable 360° reflection-type phase shifters. Linnartz et al. [99] have designed a multi-antenna rectifier circuit to improve the RF-DC rectification efficiency and directional uniformity using multiple antennas in energy-scavenging wireless sensors, which cannot afford beam steering due to energy, area and cost limitations. Tawk et al. [100] have designed the integration of arrays of inverted F-antennas on the top of a solar cell where the antenna elements are placed orthogonal to the left and right edges of the solar cell and one array composed of two element inverted-F antennas covering a specific operating bandwidth is placed on each edge. Samy et al. [101] have developed an optimization framework that exploits the RFEH from the source of RF signals with smart antenna selection schemes at the relay node for which two special case schemes, namely fixed source power antenna selection (FSP-AS) and all receive antenna selection (AR-AS), have been proposed. Zeng et al. [102] have presented a novel impedance matching method for loop antenna by integrating the in-loop ground plane (ILGP) and designed a compact Koch fractal loop for ambient RFEH applications at the GSM1800 band. Salem et al. [103] have analyzed the secrecy capacity of a half-duplex RFEH-based multi-antenna amplify-and-forward relay network in the presence of a passive eavesdropper by deriving new exact analytical expressions for the ergodic secrecy capacity for various well-known EH relaying protocols, namely, time switching relaying (TSR), power splitting relaying (PSR), and ideal relaying receiver (IRR). Cao and Li [104] have studied and compared various ambient energy sources with their advantages and disadvantages and also proposed, modeled, and simulated a new twin coil ferrite rod (TCFR) antenna for RFEH in amplitude modulation (AM) band. Ahmed et al. [105] have presented a rectifying antenna (rectenna) that operates at 2.45 GHz consisting of two-layer low-cost FR-4 substrates separated by an air gap to enhance the gain. The harmonic rejection property is embedded within the antenna design to eliminate the use of harmonic rejection filter (HRF) between the antenna and rectifier, which reduces the cost and size of the rectenna.

2.7 LOW POWER ANTENNAS

Pavone et al. [106] have designed a wideband system to recover wideband energy from electromagnetic sources present in the environment and developed a battery-free wireless sensor which is able to capture the available energy into the mentioned bandwidth. Gao et al. [107] have presented the first monolithically integrated RFEH at 71 GHz wireless temperature sensor node in 65 nm CMOS technology, containing a monopole antenna, a 71 GHz RFEH unit with storage capacitor array, an End-of-Burst monitor, a temperature sensor, and an ultra-low-power transmitter at 79 GHz to achieve a PCE of 8% for 5-dBm input power. Ramesh and Rajan [108] have presented several methods to design energy harvesting device depending on the type of energy available and also microstrip path structural rectenna for powering low consumption electrical devices. Kim et al. [109] have presented a 98-μW 457.5-MHz transmitter with output radiation power of −22 dBm, which utilizes 915-MHz wirelessly powering RF signal by frequency division using a true-single-phase-clock divider to generate the carrier frequency with very low power consumption and small die area. Yuvaraj and Samuel [110] have presented the RFEH system that can harvest energy from the ambient surroundings at the downlink RF ISM band range of the GSM-900 aimed to provide an alternative source of energy for energizing low power devices. Kadupitiya et al. [111] have investigated the power levels that can be harvested from the air and processed to achieve the energy levels that are sufficient to charge low power electronic circuits. An RF collection system has been specifically designed, constructed, and shown to successfully collect enough energy to power circuits. Chowkwale et al. [112] have presented the implementation of the low-cost RFEH System for low power remote RF sensor applications. In this implementation, conventional battery is replaced with a super capacitor which is charged by DC voltage harvested out of RF EM waves. Palandoken [113] have proposed a microstrip antenna with novel slot resonator for compact RFEH modules operating at 2.4 GHz to be utilized in DC power supply modules of modern low power wireless networks. Krishnamoothy and Umapathy [114] have presented a design and implementation of antenna for RFEH to charge low power application devices operating at three different frequency, 1800 MHz, 1900 MHz, and 2.4 GHz.

REFERENCES

1. A.M. Jie, Nasimuddin, M.F. Karim, and K.T. Chandrasekaran, "A wide-angle circularly polarized tapered-slit-patch antenna with compact rectifier for energy harvesting systems," *IEEE Antennas and Propagation Magazine*, vol. 61, no. 2, pp. 94–111, April 2019.
2. P.C. Sharma and K.C. Gupta, "Analysis and optimized design of single feed circularly polarized microstrip antennas," *IEEE Transactions on Antennas and Propagation*, vol. 31, pp. 949–955, 1983.
3. Nasimuddin, Z.N. Chen, and X. Qing, "Asymmetric-circular shaped slotted microstrip antennas for circular polarization and RFID applications," *IEEE Transactions on Antennas and Propagation*, vol. 58, no. 12, pp. 3821–3828, December 2010.

4. A.M. Jie, Nasimuddin, M.F. Karim, and K.T. Chandrasekaran, "A dual-band efficient circularly polarized rectenna for RF energy harvesting systems," *International Journal of RF and Microwave Computer-Aided Engineering*, vol. 29, pp. e21665, January 2019.
5. K.T. Chandrasekaran, Nasimuddin, A. Alphones, and M.F. Karim, "Compact circularly polarized beam-switching wireless power transfer system for ambient energy harvesting applications," *International Journal of RF and Microwave Computer-Aided Engineering*, vol. 29, pp. e21642, January 2019.
6. S.B. Vignesh, Nasimuddin, and A. Alphones, "Stubs-integrated-microstrip antenna design for wide coverage of circularly polarized radiation," *IET Microwaves, Antennas and Propagation*, vol. 11, no. 4, pp. 44–49, 2017.
7. S. Vignesh, Nasimuddin, and A. Alphones, "Circularly polarized strips integrated microstrip antenna for energy harvesting applications," *Microwave and Optical Technology Letters*, vol. 58, pp. 1044–1049, May 2016.
8. L. Bernard, Nasimuddin, and A. Alphones, "An e-shaped slotted-circular-patch antenna for circularly polarized radiation and RF energy harvesting," *Microwave and Optical Technology Letters*, vol. 58, pp. 868–875, March 2016.
9. L. Bernard, Nasimuddin, and A. Alphones, "Teo-shaped slotted-circular-patch antenna for circularly polarized radiation and RF energy harvesting," *Microwave and Optical Technology Letters*, vol. 57, no. 15, pp. 2752–2758, December 2015.
10. A.M. Jie, Nasimuddin, M.F. Karim, L. Bin, F. Chin, and M. Ong, "A proximity-coupled circularly polarized slotted-circular patch antenna for RF energy harvesting applications," *IEEE Region 10 Conference* (TENCON), Singapore, pp. 1–4, 2016.
11. Nasimuddin, K.P. Esselle, and A.K. Verma, "Wideband circularly polarized stacked microstrip antennas," *IEEE Antennas and Wireless Propagation Letters*, vol. 6, pp. 21–24, 2007.
12. Nasimuddin, Z.N. Chen, and X. Qing, "Bandwidth enhancement of a single-feed circularly polarized antenna using a metasurface," *IEEE Antennas and Propagation Magazine*, vol. 58, pp. 39–46, 2016.
13. J. Heikkinen and M. Kivikoski, "Low-profile circularly polarized rectifying antenna for wireless power transmission at 5.8 GHz," *IEEE Microwave and Wireless Components Letters*, vol. 14, pp. 162–164, April 2004.
14. X. Bao, K. Yang, O. O'Conchubhair, and M.J. Ammann, "Differentially-fed omnidirectional circularly polarized patch antenna for RF energy harvesting," *10th European Conference On Antennas and Propagation* (EuCAP), Switzerland, pp. 1–5, 2016.
15. R.A. Rahim, F. Malek, S.F.W. Anwar, S.I.S. Hassan, M.N. Junita, and H.F. Hassan, "A harmonic suppression circularly polarized patch antenna for an RF ambient energy harvesting system," *IEEE Conference on Clean Energy and Technology* (CEAT), Lankgkawi, Malaysia, pp. 1–3, 2013.
16. M.F. Ismail, M.K.A. Rahim, M.R. Hamid, H.A. Majid, and M.F.M. Yusof, "Compact dual-fed slotted circular polarization antenna with reflector for RF energy harvesting," *7th European Conference on Antennas and Propagation* (EuCAP), Gothenburg, Sweden, pp. 1–4, 2013.
17. W. Haboubi, H. Takhedmit, O. Picon, and L. Cirio, "A GSM-900 MHz and Wifi-2.45 GHz dual polarized, dual-frequency antenna dedicated to RF energy harvesting applications," *7th European Conference on Antennas and Propagation* (EuCAP), Gothenburg, Sweden, pp. 1–4, 2013.
18. S. Ghosh, S.K. Ghosh, and A. Chakrabarty, "Design of RF energy harvesting system for wireless sensor node using circularly polarized monopole antenna," *9th International Conference on Industrial and Information Systems* (ICIIS), Gwalior, India, pp. 1–4, 2014.

19. S. Ghosh, "Design and testing of RF energy harvesting module in GSM 900 band using circularly polarized antenna," *IEEE International Conference On Research in Computational Intelligence and Communication Networks* (ICRCICN), Kolkata, India, pp. 1–4, 2015.
20. H.S. Khaliq, M. Awais, W. Ahmad, and W.T. Khan, "A high gain six band frequency independent dual CP planar log periodic antenna for ambient RF energy harvesting," *Progress in Electromagnetics Research Symposium-Fall* (PIERS-FALL), Singapore, pp. 1–3, 19–22 November, 2017.
21. I.M. Adam, M.N. Yasin, P.J. Soh, K. Kamardin, M.F. Jamlos, H. Lago, and M.S. Razalli, "A simple wideband electromagnetically fed circular polarized antenna for energy harvesting," *Microwave and Optical Technology Letters*, vol. 59, pp. 2390–2397, 2017.
22. S. Shrestha, S.K. Noh, and D.Y. Choi, "Comparative study of antenna designs for RF energy harvesting," *International Journal of Antennas and Propagation*, vol. 2013, pp. 1–10, January 2013.
23. N.H. Nguyen, T.D. Bui, A.D. Le, A.D. Pham, T.T. Nguyen, Q.C. Nguyen, and M.T. Le, "A novel wideband circularly polarized antenna for RF energy harvesting in wireless sensor nodes," *International Journal of Antennas and Propagation*, vol. 2018, Article ID 1692018, pp. 9, 2018.
24. D. Sabhan, V.J. Nesamoni, and J. Thangappan, "A wide-beam, circularly polarized, three-staged, stepped-impedance, spiral antenna for direct matching to rectifier circuits," *Review of Scientific Instruments*, vol. 90, pp. 054704, 2019.
25. Y. Yang, J. Li, L. Li, Y. Liu, B. Zhang, H. Zhu, and K. Huang, "A 5.8GHz circularly polarized rectenna with harmonic suppression and rectenna array for wireless power transfer," *IEEE Antennas and Wireless Propagation Letters*, vol. 17, no. 7, pp. 1276–1280, July 2018.
26. W. Lin and R.W. Ziolkowski, "Electrically small, highly efficient, huygens circularly polarized rectenna for wireless power transfer applications," *13th European Conference on Antennas and Propagation* (EuCAP 2019), Krakow, Poland, Poland, 1–4, 2019.
27. C. Song, Y. Huang, P. Carter, J. Zhou, S. Yuan, Q. Xu, and M. Kod, "A novel six-band dual CP rectenna using improved impedance matching technique for ambient RF energy harvesting," *IEEE Transactions on Antennas and Propagation*, vol. 64, no. 7, pp. 3160–3171, July 2016.
28. S. Ladan, A.B. Guntupalli, and K. Wu, "A high-efficiency 24 GHz rectenna development towards millimeter-wave energy harvesting and wireless power transmission," *IEEE Transactions on Circuits and Systems I: Regular Papers*, vol. 61, no. 12, pp. 3358–3366, December 2014.
29. Z.W. Sim, R. Shuttleworth, and B. Grieve, "Investigation of PCB microstrip patch receiving antenna for outdoor RF energy harvesting in wireless sensor networks," *Loughborough Antennas & Propagation Conference*, Loughborough, UK, pp. 1–3, 2009.
30. B.R. Marshall and G.D. Durgin, "Staggered pattern charge collection: antenna technique to improve RF energy harvesting," *IEEE International Conference on RFID, Penang*, Malaysia, pp. 1–4, 2013.
31. A. Al-Khayari, H. Al-Khayari, S. Al-Nabhani, M. Bait-Suwailam, and Z. Nadir, "Design of an enhanced RF energy harvesting system for wireless sensors," *IEEE GCC Conference and Exhibition*, Doha, Qatar, pp. 1–4, 17–20 November, 2013.
32. C.C. Kang, S.S. Olokede, N.M. Mahyuddin, and M.F. Ain, "Radio frequency energy harvesting using circular spiral inductor antenna," *IEEE WAMICON*, Tampa, FL, USA, pp. 1–4, 2014.
33. J. Shen, J. Wen, X. Yang, X. Liu, H. Guo, C. Liu, and D. Xie, "WLAN 2.4 GHz RF energy harvester using Vivaldi antenna loaded with ZIM," *IEEE International Workshop on Electromagnetics: Applications and Student Innovation Competition* (iWEM), Nanjing, China, pp. 1–4, 2016.

34. L.J.A. Olule, G. Gnanagurunathan, N.T. Kumar, and B. Kasi, "Performance comparison of spiral resonator and EBG antennas for RF energy harvesting at 2.45 GHz ISM band," *IEEE Asia Pacific Microwave Conference* (APMC), Nanjing, China, pp. 1–3, 2017.
35. A.A. Masius, Y.C. Wong, and K.T. Lau, "Miniature high gain slot-fed rectangular dielectric resonator antenna for IoT RF energy harvesting," *International Journal of Electronics and Communications*, vol. 85, pp. 39–46, 2018.
36. S. Agrawal, R.D. Gupta, M.S. Parihar, and P.N. Kondekar, "A wideband high gain dielectric resonator antenna for RF energy harvesting application," *International Journal of Electronics and Communications* (AEU), vol. 78, pp. 24–31, 2017.
37. Y.Y. Hu, S. Sun, H. Xu, and H. Sun, "Grid-array rectenna with wide angle coverage for effectively harvesting RF energy of low power density," *IEEE Transactions on Microwave Theory and Techniques*, vol. 67, no. 1, pp. 402–413, January 2019.
38. Y.Y. Hu, H. Xu, H. Sun, and S. Sun, "A high-gain rectenna based on grid-array antenna for RF power harvesting applications," *10th Global Symposium on Millimeter-Waves*, Hong Kong, pp. 1–3, 24–26 May 2017.
39. K.K.A. Devi, S. Sadasivam, N.M. Din, C.K. Chakrabarthy, and S.K. Rajib, "Design of a wideband 377 X E-shaped patch antenna for RF energy harvesting," *Microwave and Optical Technology Letters*, vol. 54, no. 3, pp. 569–573, March 2012.
40. N.A. Zainuddin, Z. Zakaria, M.N. Husain, B. Mohd Derus, M.Z.A. Abidin Aziz, M.A. Mutalib, and M.A. Othman, "Design of wideband antenna for RF energy harvesting system," *3rd International Conference on Instrumentation, Communications, Information Technology, and Biomedical Engineering* (ICICI-BME), Bandung, pp. 1–3, 7–8 November, 2013.
41. T. Peter, T.A. Rahman, S.W. Cheung, R. Nilavalan, H.F. Abutarboush, and A. Vilches, "A novel transparent UWB antenna for photovoltaic solar panel integration and RF energy harvesting," *IEEE Transactions on Antennas and Propagation*, vol. 62, no. 4, pp. 1844–1853, April 2014.
42. G. Rai, A. Johari, and R. Shamim, "A wideband coupled E-shaped patch antenna for RF energy harvesting," *International Conference on Signal Processing and Communication* (ICSC), Noida, India, pp. 1–4, 2015.
43. H. Saghlatoon, T. Bjorninen, L. Sydanheimo, M.M. Tentzeris, and L. Ukkonen, "Inkjetprinted wideband planar monopole antenna on cardboard for RF energy-harvesting applications," *IEEE Antennas and Wireless Propagation Letters*, vol. 14, pp. 325–328, 2015.
44. M. Mrnka, Z. Raida, and J. Grosinger, "Wide-band dielectric resonator antennas for RF energy harvesting," *Conference on Microwave Techniques* (COMITE), Pardubice, Czech Republic, pp. 1–3, 2015.
45. J. Wen, D. Xie, X. Liu, H. Guo, C. Liu, X. Yang, X. Liu, and X. Yang, "Wideband collarshaped antenna for RF energy harvesting," *7th Asia Pacific International Symposium on Electromagnetic Compatibility*, Shenzhen, China, pp. 1–3, 2015.
46. J. Iwata, J. Bito, and M.M. Tentzeris, "A wideband and high gain antenna on multilayer insulation blanket for RF energy harvesting," *IEEE International Symposium on Antennas and Propagation & USNC/URSI National Radio Science Meeting*, CA, USA, pp. 1–2, 2017.
47. A. Bakytbekov, A.R. Maza, M. Nafe, and A. Shamim, "Fully inkjet printed wide band cantor fractal antenna for RF energy harvesting application," *11th European Conference on Antennas and Propagation* (EUCAP), Paris, France, pp. 1–4, 2017.
48. C.H. Lee, J.H. Park, and J.H. Lee, "Wideband tapered monopole antenna with 2 by 2 resonant loop array for electromagnetic energy harvesting and microwave power transmission," *Microwave and Optical Technology Letters*, vol. 59, no. 4, pp. 797–802, April 2017.

49. L. Vincetti, M. Maini, E. Pinotti, L. Larcher, S. Scorcioni, A. Bertacchini, D. Grossi, and A. Tacchini, "Broadband printed antenna for radiofrequency energy harvesting," *International Conference on Electromagnetics in Advanced Applications*, Cape Town, South Africa, pp. 1–4, 2012.
50. X. Wang, Z. Zhao, G. Chen, and F. He, "RF energy harvesting with broadband antenna," *IEEE Conference and Expo Transportation Electrification Asia-Pacific* (ITEC Asia-Pacific), Beijing, China, pp. 1-4, 2014.
51. A. Bertacchini, L. Larcher, M. Maini, L. Vincetti, and S. Scorcioni, "Reconfigurable RF energy harvester with customized differential PCB antenna," *Journal of Low Power Electronics and Applications*, vol. 59, pp. 257–273, 2015.
52. J. Simon, J.R. Flores Gonzalez, J.S. Gonzalez Salas, F.C. Ordaz-Salazar, and J. Flores Troncoso, "A log-periodic toothed trapezoidal antenna for RF energy harvesting," *Microwave and Optical Technology Letters*, vol. 57, no. 12, pp. 2765–2768, December 2015.
53. M. Arrawatia, M.S. Baghini, and G. Kumar, "Broadband bent triangular omnidirectional antenna for RF energy harvesting," *IEEE Antennas and Wireless Propagation Letters*, vol. 15, pp. 36–39, 2016.
54. R. Maher, E. Tammam, A.I. Galal, and H.F. Hamed, "Design of a broadband planar antenna for RF energy harvesting," *International Conference on Electrical, Electronics, and Optimization Techniques* (ICEEOT), Chennai, India, pp. 1–3, 2016.
55. C. Rivera, J.J. Pantoja, and F. Roman, "Antenna array assessment for RF energy harvesting," *IEEE International Symposium on Antennas and Propagation & USNC/URSI National Radio Science Meeting*, CA, USA, pp. 1–2, 2017.
56. X. Bai, J.W. Zhang, and L.J. Xu, "A broadband CPW fractal antenna for RF energy harvesting," *International Applied Computational Electromagnetics Society Symposium* (ACES), Suzhou, China, pp. 1–4, 2017.
57. X. Shaoa, B. Lia, N. Shahshahana, N. Goldsmana, T.S. Salterb, and G.M. Metzeb, "A planar dualband antenna design for RF energy harvesting applications," *International Semiconductor Device Research Symposium* (ISDRS), College Park, MD, 7–9 December, pp. 1–3, 2011.
58. B. Li, X. Shao, N. Shahshahan, N. Goldsman, T.S. Salter, and G.M. Metze, "Antenna-coupled dual band RF energy harvester design," *International Semiconductor Device Research Symposium* (ISDRS), College Park, MD, 7–9 December, pp. 1–3, 2011.
59. M. Arrawatia, V. Diddi, H. Kochar, M.S. Baghini, and G. Kumar, "An integrated CMOS RF energy harvester with differential microstrip antenna and on-chip charger," *25th International Conference on VLSI Design*, Hyderabad, India, pp. 1–3, 2012.
60. H.M. Saraiva, L.M. Borges, N. Barroca, J. Tavares, P.T. Gouveia, F.J. Velez, C. Loss, R. Salvado, P. Pinho, R. Goncalves, N.B. Carvalho, R. Chavez-Santiago, and I. Balasingham, "Experimental characterization of wearable antennas and circuits for RF energy harvest-ing in WBANs," *IEEE 79th Vehicular Technology Conference* (VTC Spring), Seoul, South Korea, pp. 1–3, 2014.
61. Z. Zakaria, N.A. Zainuddin, M.Z.A. Abd Aziz, M.N. Husain, and M.A. Mutalib, "A parametric study on dual-band meander line monopole antenna for RF energy harvest-ing," *Proceeding of the IEEE International Conference on RFID Technologies and Applications*, Malaysia, 4–5 September, pp. 1–3, 2013.
62. N. Barreca, H.M. Saraiva, P.T. Gouveia, J. Tavares, L.M. Borges, F.J. Velez, C. Loss, R. Salvado, P. Pinh, R. Goncalves, N.B. Carvalho, R. Chavez-Santiago, and I. Balasingham, "Antennas and circuits for ambient RF energy harvesting in wireless body area net-works," *IEEE 24th International Symposium on Personal, Indoor and Mobile Radio Communications: Fundamentals and PHY Track*, London, UK, pp. 1–3, 2013.

63. J.M. Barcak and H.P. Partal, "Efficient RF energy harvesting by using multi-band microstrip antenna arrays with multistage rectifiers," *IEEE Subthreshold Microelectronics Conference* (SubVT), MA, USA, pp. 1–4, 2012.
64. B. Li, X. Shao, N. Shahshahan, N. Goldsman, T. Salter, and G.M. Metze, "An antenna co-design dual band RF energy harvester," *IEEE Transactions on Circuits and Systems-I: Regular Papers*, vol. 60, no. 12, pp. 3256–3266, 2013.
65. M. Saad-Bin-Alam and S. Moury, "Multiple-band antenna coupled rectifier circuit for ambient RF energy harvesting for WSN," *3rd International Conference on Informatics, Electronics & Vision*, Dhaka, Bangladesh, pp. 1–4, 2014.
66. L.M. Borges, N. Barroca, H.M. Saraiva, J. Tavares, P.T. Gouveia, F.J. Velez, C. Loss, I. R.BSnaalalvsiagdhoa,m P, "Design and ev . Pinho, R. Goncalves, N.B. aluation of multi- Ca rband RF ener valho, R. Chavez gy harvesti n-Sang ctircui tiago, an ds and antennas for WSNs," *21st International Conference on Telecommunications* (ICT), Lisbon, Portugal, pp. 1–3, 2014.
67. A.A. Salih and M.S. Sharawi, "A miniaturized dual-band meander line antenna for RF energy harvesting applications," *IEEE Jordan Conference on Applied Electrical Engineering and Computing Technologies* (AEECT), Amman, Jordan, pp. 1–3, 2015.
68. L.M. Borges, R. Chavez-Santiago, N. Barroca, F.J. Velez, and I. Balasingham, "Radiofrequency energy harvesting for wearable sensors," *Healthcare Technology Letters*, vol. 2, pp. 22–27, 2015.
69. H. Kamoda, S. Kitazawa, N. Kukutsu, and K. Kobayashi, "Loop antenna over artificial magnetic conductor surface and its application to dual-band RF energy harvesting," *IEEE Trans. Antennas and Propagation*, vol. 63, no. 10, pp. 4408–4417, October 2015.
70. M. Arrawatia, M.S. Baghini, and G. Kumar, "Differential microstrip antenna for RF energy harvesting," *IEEE Transactions on Antennas and Propagation*, vol. 63, no. 4, pp. 4408–4417, April 2015.
71. Q. Zhao, J. Xu, H. Yin, Z. Lu, L. Vue, Y. Gong, Y. Wei, and W. Wang, "Dual-band antenna and high efficiency rectifier for RF energy harvesting system," *IEEE 6th International Symposium on Microwave, Antenna, Propagation, and EMC Technologies* (MAPE), Shanghai, China, pp. 1–3, 2015.
72. Z. Zhou, W. Liao, Q. Zhang, F. Han, and Y. Chen, "A multi-band fractal antenna for RF energy harvesting," *IEEE International Symposium on Antennas and Propagation* (APSURSI), Fajardo, Puerto Rico, pp. 1–2, 2016.
73. H. Pal and Y.K. Choukiker, "Design of frequency reconfigurable antenna with ambient RF-energy harvester system," *International Conference on Information Communication and Embedded System* (ICICES), Chennai, India, pp.1–3, 2016.
74. V. Hebelka, J. Velim, and Z. Raida, "Dual band Koch antenna for RF energy harvesting," *10th European Conference on Antennas and Propagation* (EuCAP), Davos, Switzerland, pp. 1–4, 2016.
75. B.A. Mouris, A.M. Soliman, T.A. Ali, I.A. Eshrah, and A. Badawi, "Efficient dual-band energy harvesting system for implantable biosensors," *17th International Symposium on Antenna Technology and Applied Electromagnetics* (ANTEM), Montreal, QC, Canada, pp. 1–4, 2016.
76. M. Mathur, A. Agarwal, G. Singh, and S.K. Bhatnagar, "The array structure of 2×2 coplanar monopole antenna with Wilkinson power combiner for RF energy harvesting application," *IEEE International Conference on Recent Advances and Innovations in Engineering* (ICRAIE-2016), Jaipur, India, pp. 1–4, December, 2016.
77. S.S. Sarma and M.J. Akhtar, "A dual band meandered printed dipole antenna for RF energy harvesting applications," *IEEE 5th Asia-Pacific Conference on Antennas and Propagation* (APCAP), Kaohsiung, Taiwan, pp. 1–3, 2016.

78. L.J. Xu, B. Huang, X. Bai, and H.P. Mao, "A dualband and broadband antenna array for ambient RF energy harvesting," *IEEE International Conference on Ubiquitous Wireless Broadband* (ICUWB), Nanjing, China, pp. 1–3, 2016.
79. V.A. Raj, "MEMS based multi-band energy harvesting for wireless sensor network applications," *International Conference on Energy Efficient Technologies for Sustainability* (ICEETS), Nagercoil, India, pp. 1–3, 2016.
80. A. Bakkali, J. Pelegri-Sebastia, T. Sogorb, V. Llario, and A. Bou-Escriva, "A dual-band antenna for RF energy harvesting systems in wireless sensor networks," *Journal of Sensors*, vol. 2016, pp. 1–8, 2016.
81. C. Loss, R. Goncalves, C. Lopes, R. Salvado, and P. Pinho, "Textile antenna for RF energy harvesting fully embedded in clothing," *10th European Conference on Antennas and Propagation* (EuCAP), avos, Switzerland, pp. 1–4, 2016.
82. A. Azam, Z. Bai, and J.S. Walling, "A low-cost, dual-band RF loop antenna and energy harvester," *IEEE Topical Conference on Wireless Sensors and Sensor Networks* (WiSNet), Phoenix, AZ, USA, pp. 1–3, 2017.
83. D.M. Elsheakh, "Planar antenna for RF energy harvesting applications," *IEEE International Symposium on Antennas and Propagation & USNC/URSI National Radio Science Meeting*, San Diego, CA, USA, pp. 1–2, 2017.
84. M.T. Hafeez and S.F. Jilani, "Novel millimeter-wave flexible antenna for RF energy harvesting," *IEEE International Symposium on Antennas and Propagation & USNC/URSI National Radio Science Meeting*, San Diego, CA, USA, pp. 1–2, 2017.
85. N.H.M. Yunus, J. Sampe, J. Yunas, and A. Pawi, "Parameter design of microstrip patch antenna operating at dual microwave-band for RF energy harvester application," *Regional Symposium on Micro and Nanoelectronics* (RSM), Batu Ferringhi, Malaysia, pp. 1–3, 2017.
86. V. Arun and L.R.K. Marx, "Micro-controlled tree shaped reconfigurable patch antenna with RF-energy harvesting," *Wireless Personal Communications*, vol. 94, pp. 2769, 2017.
87. S. Agrawal, M.S. Parihar, and P.N. Kondekar, "A quad-band antenna for multi-band radio frequency energy harvesting circuit," *International Journal of Electronics and Communications* (AEU), vol. 85, pp. 99–107, 2018.
88. S. Patil and S. Gahankari, "Design and implementation of microstrip antenna for RF energy harvesting," *International Journal of Engineering Research and Technology*, vol. 10, no. 1, pp. 487–490, 2017.
89. S.J. Darak, "Parallel aggregated MAB framework for source selection in multi-antenna RF harvesting circuit," *IEEE Wireless Communications and Networking Conference* (WCNC), Barcelona, Spain, pp. 1–4, 2017.
90. Y. Tawk, J. Costantine, F. Ayoub, and C.G. Christodoulou, "A communicating antenna array with a dual-energy harvesting functionality," *IEEE Antennas & Propagation Magazine*, vol. 60, pp. 132–144, 2018.
91. M. Kurvey and A. Kunte, "Design and optimization of stepped rectangular antenna for RF energy harvesting," *International Conference on Communication, Information & Computing Technology* (ICCICT), Mumbai, India, pp. 1–3, 2018.
92. M. Mathur, A. Agarwal, G. Singh, and S.K. Bhatnagar, "A 2×1 coplanar monopole antenna structure for wireless RF energy harvesting," *3rd International Conference on Communication Systems* (ICCS-2017), Rajasthan, India, pp. 1–3, 2017.
93. K. Celik and E. Kurt, "Design and simulation of the antenna for RF energy harvesting systems," *6th International Istanbul Smart Grids and Cities Congress and Fair* (ICSG), Istanbul, Turkey, pp. 1–3, 2018.
94. U. Muncuk, K. Alemdar, J.D. Sarode, and K.R. Chowdhury, "Multi-band ambient RF energy harvesting circuit design for enabling battery-less sensors and IoT," *IEEE Internet of Things Journal*, vol. 5, pp. 2700–2714, 2018.

95. C. Song, Y. Huang, J. Zhou, J. Zhang, S. Yuan, and P. Carter, "A high-efficiency broadband rectenna for ambient wireless energy harvesting," *IEEE Transactions on Antennas and Propagation*, vol. 63, pp. 3486–3495, 2015.
96. M. Mi, M.H. Mickle, C. Capeli, and H. Switf, "RF energy harvesting with multiple antennas in the same space," *IEEE Antennas and Propagation Magazine*, vol. 47, no. 5, pp. 100–106, October 2005.
97. D. Xie, X. Liu, H. Guo, and X. Yang, "Square electrically small EAD antenna array for RF energy harvesting from TV broadcast tower," *IEEE Asia-Pacific Microwave Conference*, Sendai, Japan, pp. 1–3, 2014.
98. S.F. Yang, T.H. Huang, C.C. Chen, C.Y. Lu, and P.J. Chung, "Beamforming power emitter design with 2×2 antenna array and phase control for microwave/RF-based energy harvesting," *IEEE Wireless Power Transfer Conference* (WPTC), Boulder, CO, USA, pp. 1–3, 2015.
99. J.P.M.G. Linnartz, Y. Wu, J.G.A. Maree, and M.K. Matters-Kammerer, "Multiple antenna rectifiers for radio frequency energy scavenging in wireless sensors," *IEEE International Symposium on Circuits and Systems* (ISCAS), Lisbon, Portugal, pp. 1–3, 2015.
100. Y. Tawk, F. Ayoub, C.G. Christodoulou, and J. Costantine, "An array of inverted-F antennas for RF energy harvesting," *IEEE International Symposium on Antennas and Propagation & USNC/URSI National Radio Science Meeting*, Vancouver, BC, Canada, pp. 1–2, 2015.
101. I. Samy, M.M. Butt, A. Mohamed, and M. Guizani, "Energy efficient antenna selection for a MIMO relay using RF energy harvesting," *IEEE Wireless Communications and Networking Conference* (WCNC), Doha, Qatar, pp. 1–3, 2016.
102. M. Zeng, A.S. Andrenko, X. Liu, H.Z. Tan, and B. Zhu, "Design of fractal loop antenna with integrated ground plane for RF energy harvesting," *International Conference on Mathematical Methods in Electromagnetic Theory*, Lviv, Ukraine, pp. 1–4, 2016.
103. A. Salem, K.A. Hamdi, and K.M. Rabie, "Physical layer security with RF energy harvesting in AF multi-antenna relaying networks," *IEEE Transactions on Communications*, vol. 64, no. 7, pp. 3025–3038, July 2016.
104. S. Cao and J. Li, "A high efficiency twin coil ferrite rod antenna for RF energy harvesting in AM band," *5th International Conference on Enterprise Systems*, Beijing, China, pp. 1–3, 2017.
105. S. Ahmed, Z. Zakaria, M.N. Husain, and A. Alhegazi, "Integrated rectifying circuit and antenna design with harmonic rejection for RF energy harvesting," *11th European Conference on Antennas and Propagation* (EUCAP), Paris, France, pp. 1–5, 2017.
106. D. Pavone, A. Buonanno, M.D. Urso, and F.D. Corte, "Design considerations for radio frequency energy harvesting devices," *Progress in Electromagnetics Research B*, vol. 45, pp. 19–35, 2012.
107. H. Gao, M.K. Matters-Kammerer, P. Harpe, D. Milosevic, U. Johannsen, A. Van Roermund, and P. Baltus, "A 71GHz RF energy harvesting tag with 8% efficiency for wireless temperature sensors in 65nm CMOS," *IEEE Radio Frequency Integrated Circuits Symposium*, Seattle, WA, USA, pp. 1–4, 2013.
108. G.P. Ramesh and A. Rajan, "Microstrip antenna designs for RF energy harvesting," *International Conference on Communication and Signal Processing*, Melmaruvathur, India, pp. 1–4, 2014.
109. Y.J. Kim, H.S. Bhamra, J. Joseph, and P.P. Irazoqui, "An ultra-low-power RF energy-harvesting transceiver for multiple-node sensor application," *IEEE Trans. Circuits and Systems—II: Express Briefs*, vol. 62, no. 11, pp. 1028–1032, November 2015.

110. K. Yuvaraj and A.A. Samuel, "A patch antenna to harvest ambient energy from multiband RF signals for low power devices," *International Journal of Emerging Technology in Computer Science & Electronics* (IJETCSE), vol. 1, pp. 1–4, 2015.
111. J.C.S. Kadupitiya, T.N. Abeythunga, P.D.M.T. Ranathunga, and D.S. De Silva, "Optimizing RF energy harvester design for low power applications by integrating multi stage voltage doubler on patch antenna," *8th International Conference on Ubi-Media Computing* (UMEDIA), Colombo, Sri Lanka, pp. 335–338, 2015.
112. B. Chowkwale, D. Yadav, and R. Abhyankar, "Energy harvesting techniques for low power RF sensors," *17th International Conference on Advanced Communication Technology* (ICACT), Seoul, South Korea, pp. 1-3, 2015.
113. M. Palandoken, "Microstrip antenna with compact anti-spiral slot resonator for 2.4 GHz energy harvesting applications," *Microwave and Optical Technology Letters*, vol. 58, no. 6, pp. 1404–1408, June 2016.
114. R. Krishnamoothy and K. Umapathy, "Design and implementation of microstrip antenna for energy harvesting charging low power devices," *4th International Conference on Advances in Electrical, Electronics, Information, Communication and Bio-Informatics* (AEEICB), Chennai, India, pp. 1–4, 2018.

3 Rectifiers for RFEH Systems

3.1 INTRODUCTION

There has been a growing interest in radio frequency energy harvesting (RFEH) as the availability of ambient radio frequency (RF) energy has increased due to advancements in broadcasting and wireless communication systems. Efficient RFEH is a very challenging issue as it deals with low levels of RF power available in the environments. Some other unpredictable issues are distance from the RF power sources, transmission medium, power density, etc. The RFEH is a process of harvesting electrical energy from RF signals emitted by dedicated, ambient, or unknown RF sources. The main design challenge of the rectenna (antenna+rectifier) is its power conversion efficiency (PCE), which is the measure of the efficiency of a rectifier in converting received RF energy into DC current. PDC denotes the output DC power, and PRF denotes the input RF power. The voltage at the input of the rectifier varies according to the frequency of the incident wave, which results in variation in the diode impedance leading to a reduced PCE due to impedance mismatch.

For an efficient RFEH system, an efficient rectifier circuit is important to improve the RF-to-DC conversion efficiency. The overall efficiency of the RFEH system depends on RF-to-DC conversion efficiency of the rectifier and receiving antenna design. The rectifier circuit consists of an impedance matching network and rectifier. Different types of rectifier circuits have been used in RFEH systems to enhance the overall efficiency of the system. The rectifier contains four parts: a matching network, an RF-to-DC converter, a DC-pass filter, and a load. A matching circuit is key element of the rectifier to transfer RF-received power to DC converter. The detailed description of matching circuit is provided in Chapter 8.

3.2 AN OVERVIEW OF MATCHING NETWORKS

The energy inefficiency of an RFEH system is mainly due to power leakage during transmission and mismatching between the receiver antenna and rectifier circuits. An impedance matching circuit enables maximum power transfer between antenna and load. The rectifier and remaining circuit are considered as load in the RFEH system. When there is an impedance mismatch, the incident wave gets reflected at the load, which leads to reduction in efficiency. A matching circuit ensures identical impedance between the source and the load. It can also act as a low pass filter (LPF) to reject higher-order harmonics generated by the rectifying circuit which can be re-radiated by the antenna creating loss [1,2]. Hence, a matched filter/impedance circuit is desired between the rectifier and the antenna. An impedance variation of the rectifier with variation in input power as well as the load leads to degradation in PCE. The preferred characteristic of the matching circuit is that it should be able to match the load impedance with the antenna impedance at different frequencies within wideband, load resistance, and input power. It should have a small form factor as well as wideband of operation. The main design challenge is that the antenna impedance changes with load and input power. Tuning circuits can be used to change impedance to a desired value. Bandwidth improvement is provided by the second-order matching circuit than the first-order one [3]. But bandwidth decreases rapidly if the order is increased beyond two matching circuit stages. Lumped elements and distributed microstrip lines are used for implementing matching circuit. Compared to distributed line, lumped element-based matching circuit has lower Q offering wider bandwidth. Due to the parasitic effects associated with lumped elements, these are not preferred at higher frequencies. The lumped circuits used for matching are commonly T-network, π-network, shunt inductor, L-network, gamma matching network, band pass filter (BPF), etc. Commonly used matching network for low-power regimes is the simple two-component L-type matching network shown in Figure 3.1 as it provides impedance matching with a minimal loss [4] and an inductor also pre-boosts the input signal reaching the rectifier. Due to high-quality factor (Q), these matching networks generally have a narrow bandwidth. L_{mat} and C_{mat} of the matching network are designed with a source impedance of 50-Ω as shown in the figure. Some matching networks and rectifier circuits have been reviewed in detail in Ref. [5].

Source resistance can be calculated as follows:

$$R_S = R_{\text{in}}\left(\frac{1}{1+Q^2}\right) \tag{3.1}$$

The quality factor, Q, can be defined as follows:

$$Q = \sqrt{\frac{R_{\text{in}}}{R_S} - 1} \tag{3.2}$$

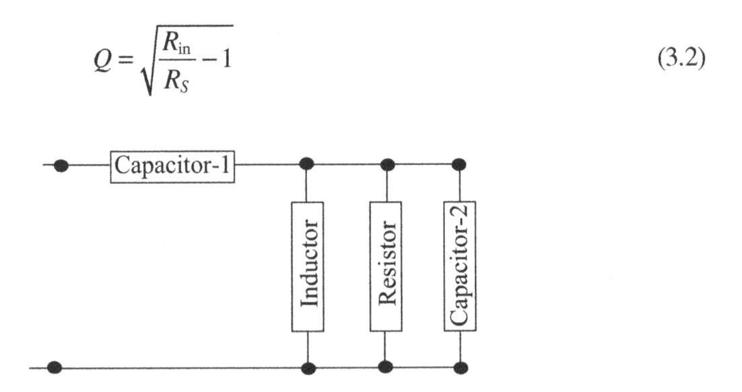

FIGURE 3.1 L-Type matching network. (From Divakaran, S.K., Krishna, D.D., and Nasimuddin, *Int. J. RF Micro. Computer-Aided Eng.*, 29, 1–15, 2019. With permission.)

The quality factor expressed in terms of imaginary part of impedance is as follows:

$$Q = \frac{Im(Z)}{Re(Z)} \approx \frac{R_{\text{in}}}{\omega_0 L_{\text{mat}}} - \omega_0 C_{in} R_{\text{in}} \tag{3.3}$$

L_{mat} of matching network can be calculated as follows:

$$L_{\text{mat}} = \frac{R_{\text{in}}}{\omega_0 \left(Q + \omega_0 C_{\text{in}} R_{\text{in}}\right)} \tag{3.4}$$

C_{mat} is calculated by equating imaginary part to zero as follows:

$$C_{\text{mat}} = \frac{R_{\text{in}}}{L_{\text{mat}}\left(R_{\text{in}} - \text{R}_S\right)} \frac{1}{\left(\omega_0^2 - \frac{1}{L_{\text{mat}} C_{\text{in}}}\right)} \tag{3.5}$$

An L-type matching network provides restriction on tuning of two components. This can be overcome by adding an additional inductor to the circuit transforming it to T- or π-type network. The π-type matching network is shown in Figure 3.2 with a diode replaced by equivalent impedance, and these are superior to L-type networks since they provide an extra degree of freedom with a greater amplitude of resonance. The output voltage varies rapidly with frequency compared to the L-type as shown in Figure 3.3a because of the presence of frequency-dependent element.

The design for π-type network is given below:

$$Z_{\text{in}} = \left\{\left[\left(R_L - jX_L\right) \| \left(\frac{1}{j\omega C_2}\right)\right] + j\omega L\right\} \| \left(\frac{1}{j\omega C_1}\right) \tag{3.6}$$

The parasitic losses associated with passive elements increase with frequency, and capacitor behavior will change into inductance as frequency increases.

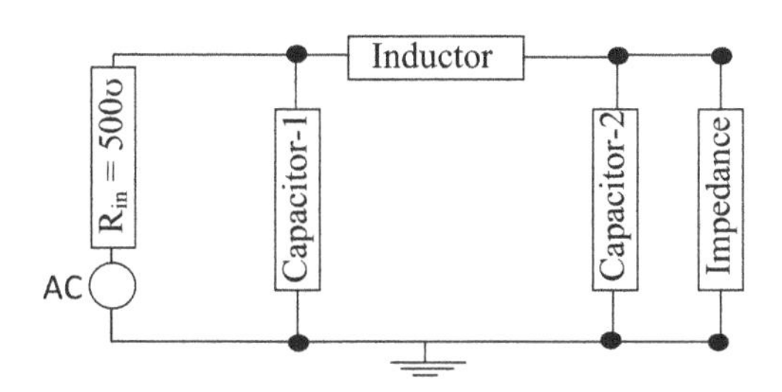

FIGURE 3.2 π-Type matching network. (From Divakaran, S.K., Krishna, D.D., and Nasimuddin, *Int. J. RF Micro. Computer-Aided Eng.*, 29, 1–15, 2019. With permission.)

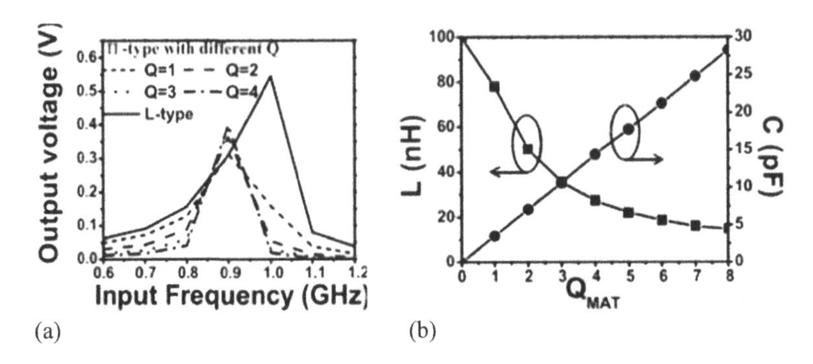

FIGURE 3.3 (a) Output voltage as the function of frequency and (b) variations of C and L with Q. (From Divakaran, S.K., Krishna, D.D., and Nasimuddin, *Int. J. RF Micro. Computer-Aided Eng*., 29, 1–15, 2019. With permission.)

From Figure 3.3b, it can be observed that L value changes drastically as Q increases. So, higher Q circuits can be easily implemented due to low L values.

The theoretical limitation of impedance matching bandwidth to parallel load impedance is provided by Bode [6] and Fano [7]:

$$\int_0^\infty \ln \frac{1}{|\Gamma(\omega)|} d\omega < \frac{\pi}{R_{\text{load}} C_{\text{load}}} \tag{3.7}$$

For series load,

$$\int_0^\infty \frac{1}{\omega^2} \ln \frac{1}{|\Gamma(\omega)|} d\omega < \pi R_{\text{load}} C_{\text{load}} \tag{3.8}$$

where Γ is a reflection coefficient. Γ is ideally 0 for the designed bandwidth range (Δf) and 1 for outside this bandwidth. Hence,

$$\Gamma\omega \ln \frac{1}{|\Gamma(\omega)|_{\min}} < \frac{\pi}{R_{\text{load}} C_{\text{load}}} \tag{3.9}$$

$$\Gamma = e^{-\frac{1}{2\Delta f R_{\text{load}} C_{\text{load}}}} \tag{3.10}$$

Thus, we find, according to Bode and Fano, that efficient matching is achieved at the cost of bandwidth. Table 3.1 shows comparison of some matching networks. The T- and π-type matching networks are almost the same.

Broadband impedance matching circuits with two branching sections has been presented in Refs. [8,9]. Radial stub, short stub, and a 6 nH chip inductor are incorporated in the upper branch to obtain impedance matching around the frequency

TABLE 3.1
Comparison of Different Matching Networks

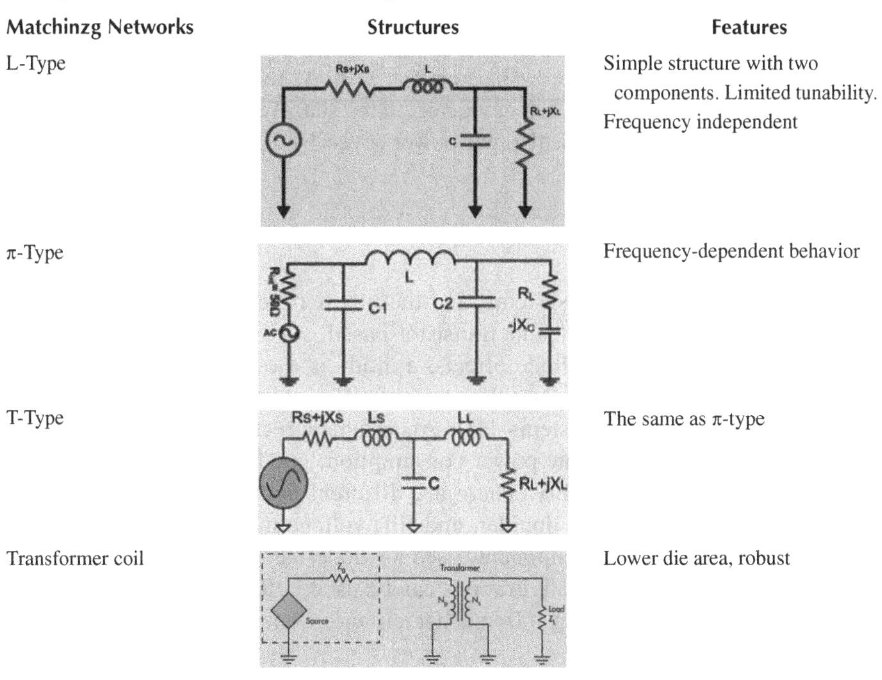

Matchinzg Networks	Structures	Features
L-Type		Simple structure with two components. Limited tunability. Frequency independent
π-Type		Frequency-dependent behavior
T-Type		The same as π-type
Transformer coil		Lower die area, robust

range of 1.8–2.5 GHz. The compact rectennas can also be achieved by eliminating matching networks. Song et al. [9] have developed a rectenna system without matching network by changing the antenna to a high-impedance antenna that can directly conjugate match with the specific rectifier impedance. The antenna is an off-center-fed dipole antenna operating at frequency band 1.8–2.5 GHz with imaginary part of impedance varying between 0 and 300 Ω in the desired band. The rectenna was able to achieve a conversion efficiency of 75% at 0 dBm input power, which is high compared to other broadband rectennas. RC LPF is incorporated at the output of the rectifier circuits to ensure pure DC reaches the load. Moreover, the radial stubs are used to match load variations, and stub-based circuits are implemented to reduce parasitic losses at high frequencies.

The RFEH circuits are characterized by two parameters: sensitivity and efficiency. The efficiency, expressed as PCE, is the measure of ability of a rectifier to convert the incoming RF energy into DC current. PCE of a diode changes with input power variations. The RF-to-DC conversion efficiency of a rectenna can be calculated as follows:

$$\text{RF to DC conversion efficiency} = \frac{\text{output DC power}}{\text{input RF power}} \tag{3.11}$$

The voltage at the input of the rectifier varies according to frequency resulting in variation in the diode impedance, thus leading to efficiency degradation due to mismatch.

In low-power regions, the efficiency is less because the forward drop of the diode is greater or comparable to the voltage swing. The PCE is also affected with the generation of higher-order harmonics [3,10]. As the voltage swing exceeds breakdown voltage, V_{br}, the efficiency deteriorates sharply. Breakdown voltage of the diode determines the critical input power given by $V_{br}^2/4R_L$, where R_L is the rectenna load resistance.

3.3 RECTIFIERS

Rectifier is a key part in EH systems, and the rectifier technologies fall into two main categories: diode-based and transistor-based. As transistor-based rectifiers are developed based on diode effects, a diode is the key component of any energy converter and can be found in various EH systems such as solar, wind, thermal, kinetic, and RF systems [11]. An efficient rectifier configuration for RFEH system must have a low power consumption, good power sensitivity, and good power handling capability. There are different configurations of rectifier: (i) basic rectifier, (ii) voltage doubler, and (iii) voltage multiplier. The rectifiers are classified based on the components used as diode-based and MOSFET-based. Different types of rectifier configurations can be used in RFEH system, and these include Dickson charge pump, Greinacher circuit, Compression networks, and Villard circuit.

3.3.1 Diode-Based Rectifiers

Diode-based rectifier circuits are most commonly used due to their low forward voltage drop compared to CMOS circuits. Single-stage voltage doubler-based rectifier circuit is widely used for high- and medium-power applications. Schottky barrier diodes are commonly used in rectenna applications [12,13]. Diode with a lower forward voltage is the best choice since it can achieve higher PCE. High efficiency and high output power at low input voltage makes single-stage voltage doubler rectifier a good choice for RFEH system. Figure 3.4 shows that for achieving high efficiency at low input power, low turn-on threshold is needed. However, low threshold voltage lowers the breakdown voltage. The level of output power can be increased to a maximum with a large reverse breakdown voltage. PCE depends on zero bias diode junction capacitance (C_{j0}), diode breakdown voltage (V_{br}), series resistance (R_s), switching speed of the diode, and a low threshold voltage. It is also affected by losses in the substrate and the transmission lines (Figure 3.5).

Single-stage voltage doubler circuit is discussed in Ref. [14]. During negative half cycle, the diode D_2 is forward-biased and it acts as a short circuit. Thus, no current flows through D_1 and energy is stored in C_2. Series diode D_1 rectifies the positive half

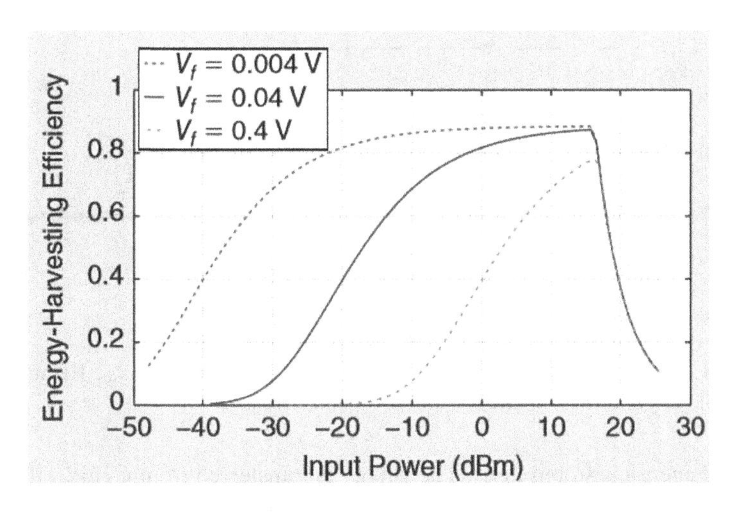

FIGURE 3.4 Performance variation. (From Sze, S., *Semiconductor Devices: Physics and Technology*, Wiley, New York, 2002. With permission.)

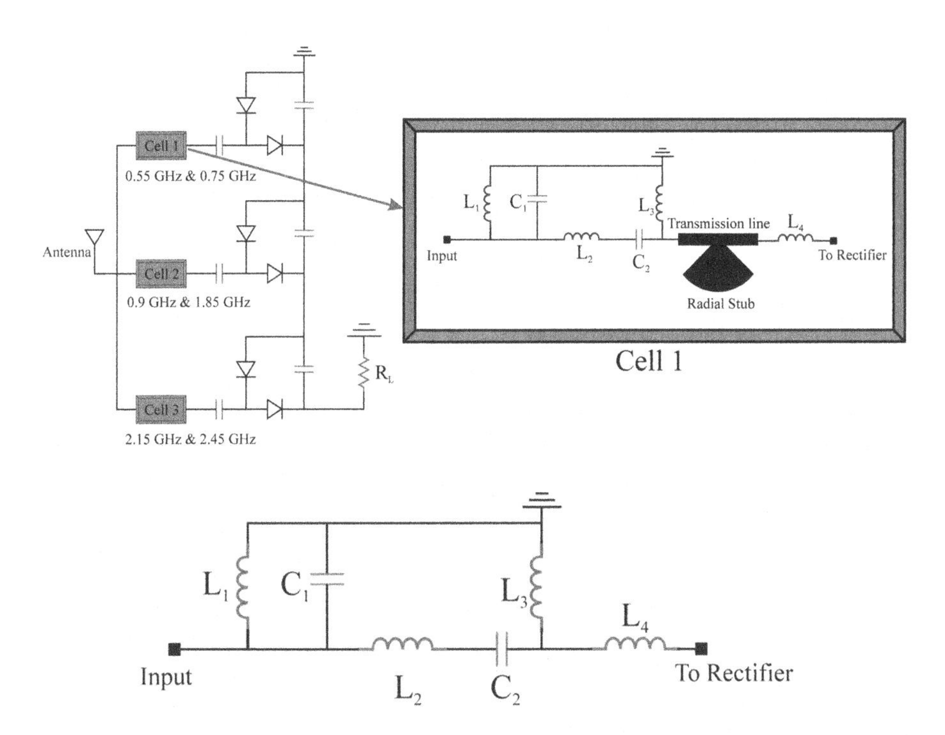

FIGURE 3.5 Novel six-band dual CP rectenna. (From Song, C., Huang, Y., Carter, P., Zhou, J., Yuan, S., Xu, Q., and Kod, M., *IEEE Trans. Antennas Propag.*, 64, 3160–3171, 2016. With permission.)

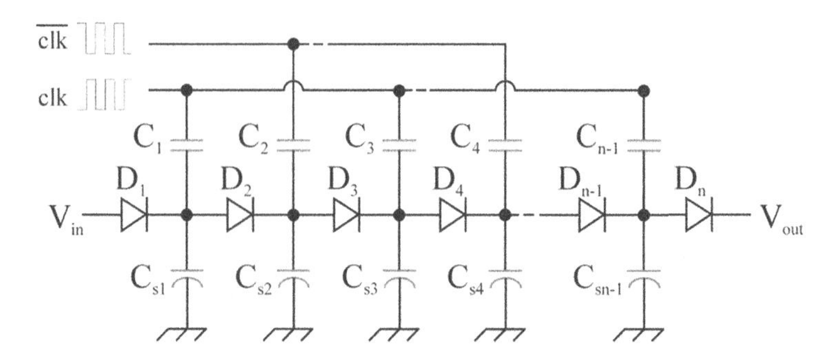

FIGURE 3.6 Dickson charge pump. (From Toudeshki, A., Mariun, N., Hizam, H., and Wahab, N.I.A., *Chinese J. Eng.*, 2014, 6, 2014. With permission.)

cycle, and energy is stored in C_1. The energy is transferred from C_2 to C_1 during the next period and discharged to R_L. The output voltage is the difference between twice the input voltage and diode drop. An overview of the progress achieved in an RFEH system and rectifier circuits is provided in Ref. [15] with some modified forms of existing CMOS-based voltage doubler circuits.

The Cockroft–Walton voltage doubler could produce an output voltage that is three times greater than the applied voltage [16]. The main disadvantage is the high coupling voltage drop, which results in lower gain for the circuit. Since the output capacitors are holding a floating charge, it is very difficult to store charge individually for other applications. Another type of charge pump is the Dickson charge pump shown in Figure 3.6. It requires clock pulses as input for capacitors and is suitable for low-voltage applications. The main drawback is the requirement of clock pulses for capacitors, which limits its application in high-voltage cases.

3.3.2 MOSFET-Based Rectifiers

Limitation of diodes can be overcome by the MOSFET(metal oxide semiconductor field effect transistor) technology. A major advantage of this technology is the fast switching speed, but it is susceptible to thermal runaway and electromagnetic interference (EMI). It also requires a high threshold voltage, which limits the efficiency of EH circuits [17]. Many compensation methods are used to overcome this disadvantage. The Dickson charge pump is also incorporated in MOSFETs. The voltage loss across MOSFET devices leads to low efficiency. This is further deteriorated by reverse leakage current. Another major disadvantage of the MOSFET-based circuits is that as frequency increases, efficiency decreases due to increased power loss in the MOSFET occurring from the reverse leakage current [18]. Harmonics and intermodulation products are produced by diodes due to its non-linear behavior reducing PCE. Increased incident power level reduces efficiency due to increase in parasitic losses with harmonic generation. So, there is a trade-off between all these. For low-power handling applications, low threshold voltage diodes are preferred, while high reverse breakdown voltage diodes are preferred for high-power applications.

One method for threshold compensation is the cross-coupled technique [19] shown in Figure 3.7. It allows for simultaneous function of charging and discharging phases through control gate circuit. Higher efficiency is achieved through threshold voltage compensation. The main disadvantage is that to achieve high efficiency, a large number of stages are required, which makes the circuit more complicated and bulkier. In Ref. [20], a rectifier is designed to power RFID tags (Figure 3.8). The RF signals are

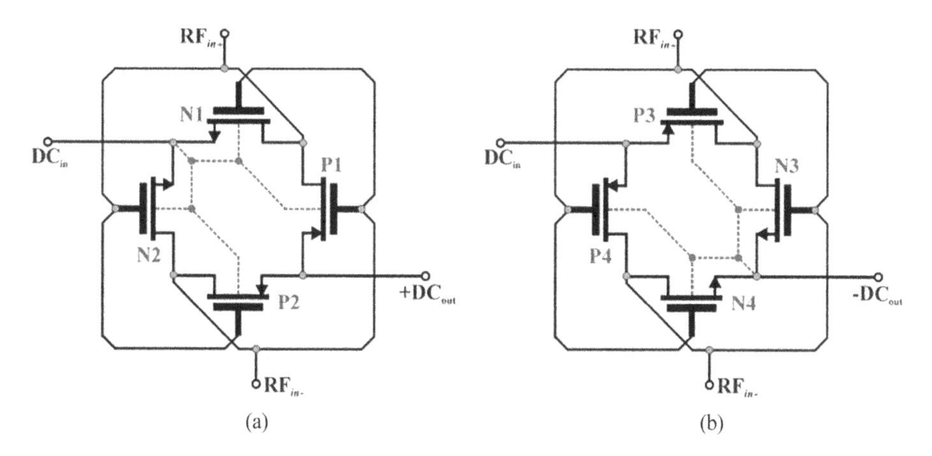

FIGURE 3.7 Circuit configuration of lower DC feeding (LDCF) body biasing technique for DDCC rectifier: (a) positive rectifier and (b) negative rectifier. (From Moghaddam, A.K., Chuah, J.H., Ramiah, H., Ahmadian, J., Mak, P.I., and Martins, R.P., *IEEE Trans. Circuits Sys. I Reg. Papers*, 64, 992–1002, 2017. With permission.)

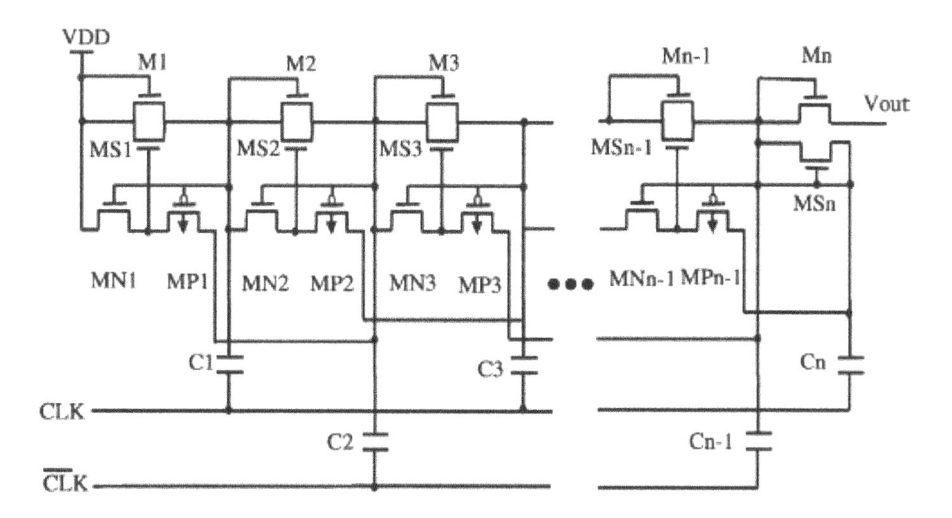

FIGURE 3.8 Cascaded cross-coupled voltage multiplier. (From Moisiais, Y., Bouras, I., and Arapoyanni, A., *VLSI Design*, 15, 477–483, 2002. With permission.)

TABLE 3.2
Comparison of Various Rectifier Topologies for RFEH System

Ref.	Rectifiers	Frequency (GHz)	OLR (kΩ)	Output Voltage (V)	Efficiency
[23]	Schottky diode	2.45	1.4	1 V at Pin = −3.2 dbm	50% at Pin −17.2 dBm
[29]	Diode-based Greinacher voltage multiplier for multiband	0.9, 1.8, 2.1, 2.45	11	0.9 at Pin = −15 dBm	15% at Pin = −20 dBm
[26]	Single-stage, diode-based voltage multiplier	2.45	NR*	5.5	75% for Pin = 19 dBm
[27]	VD-type rectifier	2.45	1	6.1	55% for $R_L = 1$ K
[28]	Half-wave rectifier for low input power	2.45	NA	NA	61.4% at −5 dBm
[30]	Diode-connected, zero-threshold NMOS transistor	2.2	5000	1 V at −25.5 dBm	NA
[32]	Schottky diode-based rectifier	2.45	2.8	3 at Pin = −15 dBm	45% at Pin = 0 dBm
[33]	Schottky diode-based rectifier	2–18	0.1	NR[a]	20% at Pin = 0.1 mW/cm^2
[34]	Full-wave Greinacher circuit	1.8	12	1.8 V at Pin = 10 $\mu W/cm^2$	60% at Pin = 10 $\mu W/cm^2$
[35]	0.18 µm CMOS	0.95	1	0.4	NA
[36]	0.18 µm differential CMOS	0.953	10	1.8	67.5% at −12.5 dBm
[37]	Bridge rectifier (HSMS2820)	0.9–2.45	2.4	6.5	80% at 23 dBm
[38]	HSMS 285C	0.945	0.1	2.2–4.5	52% at −10 dBm
[39]	Sub-rectifier circuits with HSMS 2822 and HSMS 2860	2.45	0.82 and 1.2	NA	50% at 0–24 dBm

Source: Divakaran, S.K., Krishna, D.D., and Nasimuddin, *Int. J. RF Micro. Computer-Aided Eng.*, 29, 1–15, 2019. With permission.

[a] NR—not recorded.

passed from the tag antenna into an impedance matching network. It uses a half wave rectifier of '*n*' stages. The incoming power is passed into the next stage only during one half cycle. The system works well at a distance of 3 m from the source for a short duration. Its main disadvantage is that as the number of stages increases, the size of the circuit increases, which leads to efficiency degradation. The system works well only for high input power level applications.

An effective way to improve the efficiency of the entire system is to use a passive network to boost the voltage amplitude at the input of the charge-pump circuit [21]. Buck boost converter is used in between antenna and rectifying circuit to improve efficiency. Input impedance can be made independent of load resistance and input power by using a converter controlled by pulse oscillator circuit. Inductor, diode, switching elements, and capacitors are the main components used in discontinuous conduction mode to have stable input impedance. If energy is harvested using multiple antennas or rectennas, a DC combining circuit must be used to get an overall DC output voltage.

3.3.3 Comparison of Various Rectifier Topologies

Various rectifier topologies [22–39] are compared in Table 3.2. Based on the comparison, it can be concluded that diode-based rectifiers provide better efficiency compared to the MOSFET-based rectifiers. Most of the rectifier implementations are based on the CMOS technology. Even though it can work with lower RF voltage compared to HSMS [40] and SMS [41] technologies, efficiency is inferior compared to other technologies. Using HSMS and SMS technologies, efficiency above 40% can be achieved at –20 dBm input power. Similarly, a single-stage voltage rectifier circuit provides a greater efficiency than multi-stage rectifiers because of less parasitic losses.

Recently, a passive RF-to-DC converter (Figure 3.9) for RFEH at ultra-low input power at 868 MHz has been presented in Ref. [42], which consists of a reactive

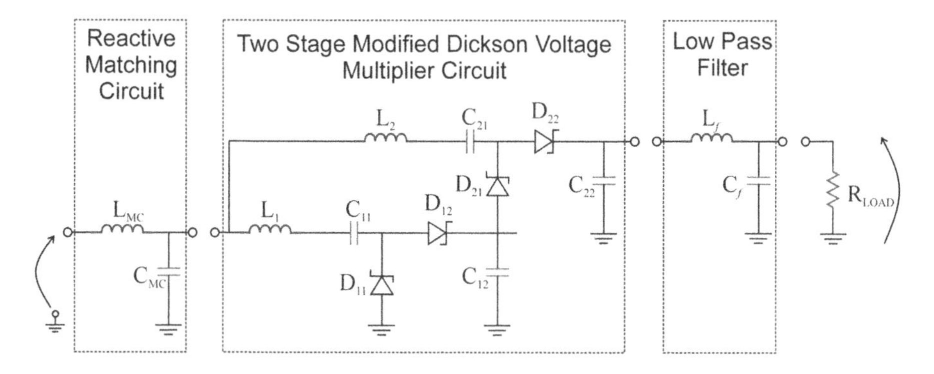

FIGURE 3.9 Efficient RF-DC converter circuits for low RFEH. (From Chaour, I., Fakhfakh, A., and Kanoun, O., *Sensors*, 17, 1–14, 2017. With permission.)

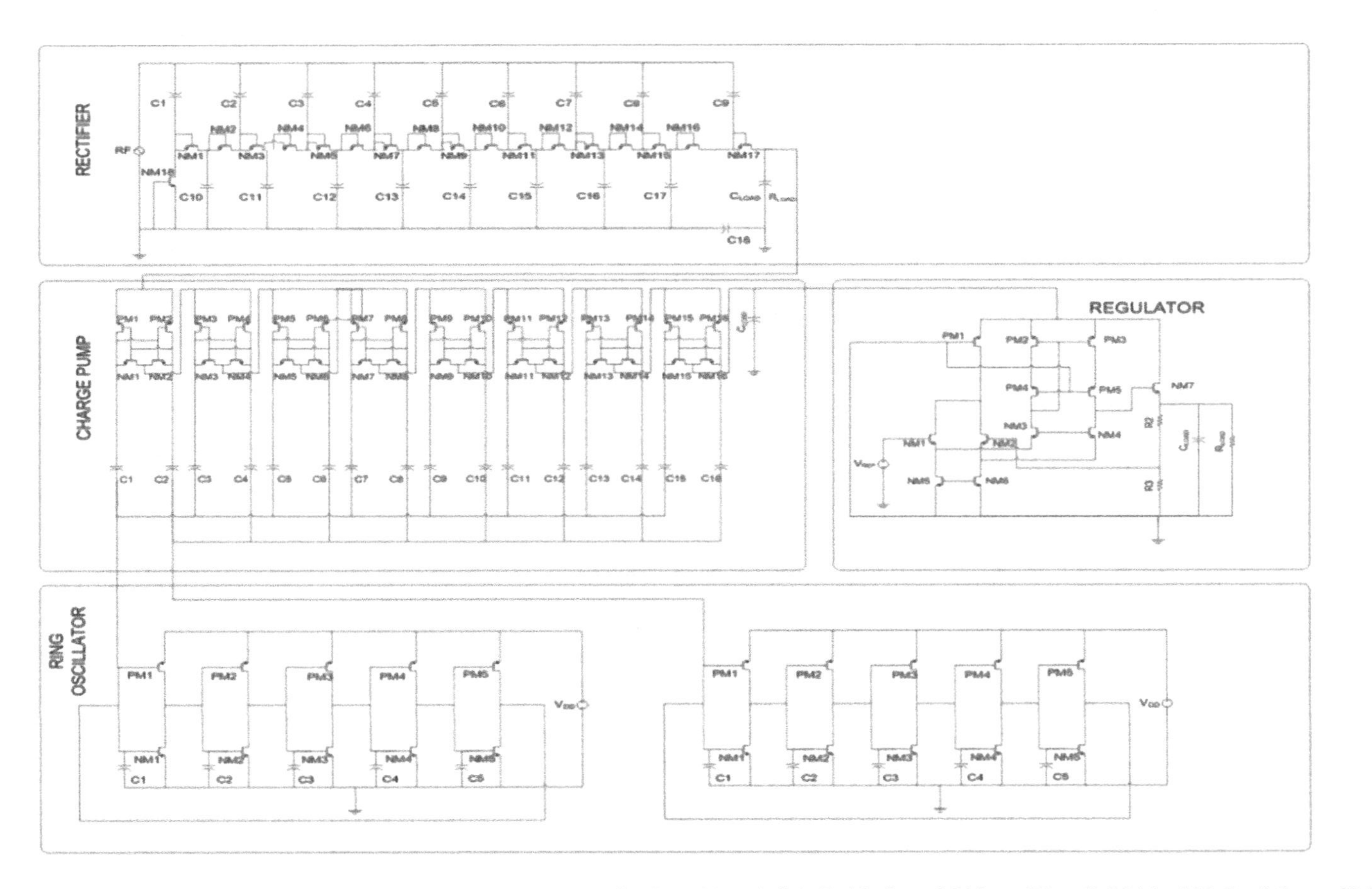

FIGURE 3.10 Multi-stage RF-to-DC conversion circuit. (From Rosli, M.A., Murad, S.A.Z., Norizan, M.N., and Ramli, M.M., *AIP Conf. Proc.*, 2045, 020089, 2018. With permission.)

matching circuit and a combined voltage multiplier and rectifier. The stored energy in the input inductor and capacitance during the negative wave is conveyed to the output capacitance during the positive one, so it is capable of harvesting low ambient RF power levels using a novel multiplier circuit technique and high-quality components to reduce parasitic effects and threshold voltages. In Ref. [43], a multi-stage NMOS rectifier with a cross-coupled charge pump has been used for RF-to-DC conversion in RFEH system (Figures 3.10 and 3.11).

In Ref. [44], several selected designs of multiband and broadband rectifiers have been compared in Table 3.3, and finally, the most optimum power combining mechanisms that can be used when dealing with multi-element rectifier have been demonstrated.

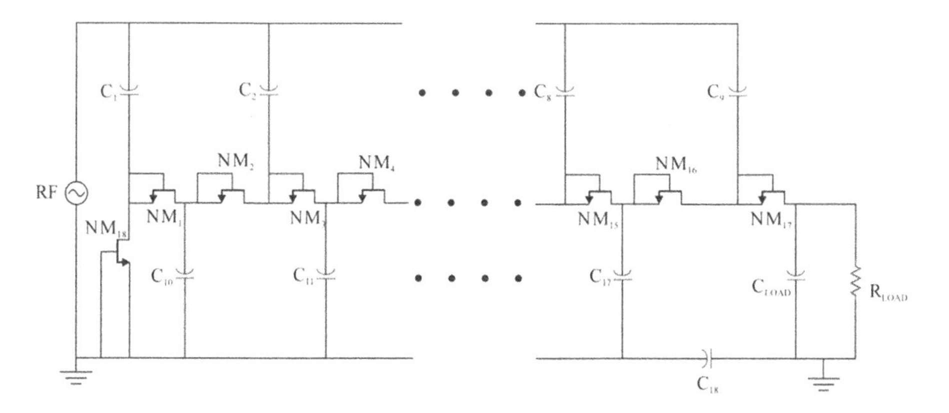

FIGURE 3.11 Multi-stage RF-to-DC conversion circuit. (From Rosli, M.A., Murad, S.A.Z., Norizan, M.N., and Ramli, M.M., *AIP Conf. Proc.*, 2045, 020089, 2018. With permission.)

TABLE 3.3
Comparison of Different Multi-Band Rectifier Designs

Ref.	Operating Frequency	Topology	P_{av} (dBm)	Efficiency	R_L (Ω)	Vdc	No. of Elements
[45]	0.9, 1.8, 2.4 GHz	Full-wave rectifier	1.76 dBm	60% at 0.9 GHz 47% at 1.76 GHz 33.5% at 2.45 GHz	6300	2.381 V at 900 MHz 2.1 V at 1.76 GHz 1.775 V at 2.45 GHz	1
[46]	0.9, 1.8, 2.4 GHz	Three charge pump branches	–15 dBm	% at 0.9 GHz % at 1.8 GHz % at 2.45 GHz	50,000	0.82 V at 0.9 GHz 0.85 V at 1.8 GHz 0.62 V at 2.45 GHz	1
[47]	0.915, 2.45 GHz	Single-diode rectifier	–11 dBm at 0.915 GHz –13.5 dBm at 2.45 GHz	47% at 0.915 GHz 56.2% at 2.45 GHz	2200	0.2 V at 0.915 GHz 0.3135 V at 2.45 GHz	1
[48]	0.9, 1.8, 2.4 GHz	Single-series diode	27 dBm	~50% at 0.9, 1.8, 2.4 GHz	50	~0.25 V at 0.9, 1.8, 2.4 GHz	1
[49]	0.85, 1.85 GHz	Single-series diode	–20 dBm	15% at 0.85 GHz 15% at 1.85 GHz	2200	57.4 at 0.85 GHz 57.4 at 1.85 GHz	1
[49]	0.85, 2.45 GHz	Single-series diode	–20 dBm	18% at 0.85 GHz 10% at 2.45 GHz	2200	46.9 at 0.85 GHz 62.9 at 2.45 GHz	1
[50]	0.791, 1.57, 2.34 GHz	Single-series diode	–10 dBm	~40% at 0.791, 1.57, 2.34 GHz	1470	~242 at 0.791, 1.57, 2.34 GHz	1
[51]	2.41–2.47 GHz	Single-shunt diode	–20 dBm	24.3	1800	70	1
[52]	~0.8–1 GHz	Five-stage charge pump circuit	18 dBm	31.8	200	6000	1
[53]	0.6–1.15 GHz	Class F-1 amplifier based	40 dBm	>60	34	521.5	1
[34]	2–18 GHz	Single diode	[–17…+15] at 3GHz	[0.1–20] at 3GHz	100	[2.5–790] at 3 GHz	64
[54,55]	0.47–0.86 GHz	Single diode	10	>60			1

Source: Collado, A., Daskalakis, S.N., Niotaki, K., Martinez, R., Bolos, F., and Georgiadis, A., *Radioengineering*, 26, 1–7, 2017. With permission.

REFERENCES

1. L. Marnat, M.H. Ouda, M. Arsalan, K. Salama, and A. Shamim, "On chip implantable antennas for wireless power and data transfer in a glaucoma-monitoring SoC," *IEEE Antennas and Wireless Propagation Letters*, vol. 11, pp. 1671–1674, 2012.
2. A.R. Lopez, "Review of narrowband impedance-matching limitations," *IEEE Antennas and Propagation Magazine*, vol. 46, no. 4, pp. 88–90, August 2004.
3. T.W. Yoo and K. Chang, "Theoretical and experimental development of 10 and 35 GHz rectennas," *IEEE Transactions on Microwave Theory and Techniques*, vol. 40, no. 6, pp. 1259–1266, June 1992.
4. Y. Sun and J. Fidler, "Design of impedance matching networks," *IEEE International Symposium on Circuits and Systems*, London, UK, pp. 1–4, May 1994.
5. S.K. Divakaran, D.D. Krishna, and Nasimuddin, "RF energy harvesting systems: an overview and design issues," *International Journal of RF and Microwave Computer-Aided Engineering*, vol. 29, no. 1, pp. 1–15, January 2019.
6. W.H. Bode, *Network Analysis and Feedback Amplifier Design*, 1st Edition, Princeton, NJ: Van Nostrand, 1945.
7. R.M. Fano, "Theoretical limitations on the broadband matching of arbitrary impedances," D.Sc. dissertation, Dep. of Elect. Eng., Massachusetts Inst. Technol. (MIT), Cambridge, MA, USA, 1947.
8. C. Song, Y. Huang, J. Zhou, J. Zhang, S. Yuan, and P. Carter, "A high efficiency broadband rectenna for ambient wireless energy harvesting," *IEEE Transactions on Antennas and Propagation*, vol. 63, no. 8, pp. 3486–3495, August 2015.
9. C. Song, Y. Huang, J. Zhou, P. Carter, S. Yuan, Q. Xu, and Z. Fei, "Matching network elimination in broadband rectennas for high efficiency wireless power transfer and energy harvesting," *IEEE Transactions on Industrial Electronics*, vol. 64, no. 6, pp. 3950–3961, May 2017.
10. R.K. Yadav, S. Das, and R.L. Yadava, "Rectennas design development and applications," *International Journal of Engineering, Science and Technology*, vol. 3, no. 10, pp. 7823–7841, October 2014.
11. S. Hemour and W. Ke, "Radio-frequency rectifier for electromagnetic energy harvesting: development path and future outlook," *Proceedings of the IEEE*, vol. 102, pp. 1667–1691, 2014.
12. W. Saeed, N. Shoaib, H.M. Cheema, and M.U. Khan, "RF energy harvesting for ubiquitous, zero power wireless sensors," *International Journal of Antennas and Propagation*, vol. 2018, pp. 16, Article ID 8903139, 2018.
13. S. Sze, *Semiconductor Devices: Physics and Technology*, New York: Wiley, 2002.
14. C. Song, Y. Huang, P. Carter, J. Zhou, S. Yuan, Q. Xu, and M. Kod, "A novel six-band dual CP rectenna using improved impedance matching technique for ambient RF energy harvesting," *IEEE Transactions on Antennas Propagation*, vol. 64, no. 7, pp. 3160–3171, July 2016.
15. H. Jabbar, Y.S. Song, and T.T. Jeong, "RF energy harvesting system and circuits for charging of mobile devices," *IEEE Transactions on Consumer Electronics*, vol. 56, no. 1, pp. 247–253, February 2010.
16. A. Toudeshki, N. Mariun, H. Hizam, and N.I.A. Wahab, "Development of a new cascade voltage-doubler for voltage multiplication," *Chinese Journal of Engineering*, vol. 2014, pp. 6, Article ID 948586, 2014.
17. A.C. Kailuke, P. Agarwal, and R.V. Kshirsagar, "Design and implementation of low power Dickson charge pump in 0.18m CMOS process," *International Journal of Scientific & Engineering Research*, vol. 4, pp. 1941–1944, August 2013.

18. S.S. Hashemi, M. Sawan, and Y. Savaria, "A high-efficiency low-voltage CMOS rectifier for harvesting energy in implantable devices," *IEEE Transactions on Biomedical Circuits and Systems*, vol. 6, pp. 326–335, 2012.
19. A.K. Moghaddam, J.H. Chuah, H. Ramiah, J. Ahmadian, P.I. Mak, and R.P. Martins, "A 73.9%-efficiency CMOS rectifier using a lower DC feeding (LDCF) Self-body-biasing technique for far-field RF energy-harvesting systems," *IEEE Transactions on Circuits and Systems–I: Regular Papers*, vol. 64, pp. 992–1002, 2017.
20. Y. Moisiais, I. Bouras, and A. Arapoyanni, "Charge pump circuits for low voltage applications," *VLSI Design*, vol. 15, no. 1, pp. 477–483, 2002.
21. A.J. Murley, "Device and method for harvesting, collecting or capturing and storing ambient energy," US Patent, US20150048682, February 2015.
22. H. Sun, Y.X. Guo, M. He, and Z. Zhong, "Design of a high-efficiency 2.45GHz rectenna for low-input-power energy harvesting," *IEEE Antennas and Wireless Propagation Letters*, vol. 11, pp. 929–932, 2012.
23. P. Sample, D.J. Yeage, J.R. Smith, P.S. Powledge, and A.V. Mamishev, "Energy harvesting in RFID system," *IEEE International Conference on Actual Problems of Electron Devices Engineering*, Russia, pp. 1–3, September 2006.
24. K.Y. Lin, T.K.K. Tsang, M. Sawan, and M.N.E. Gamal, "Radio-triggered solar and RF power scavenging and management for ultra-low power wireless medical applications," *IEEE International Symposium on Circuits and Systems*, Greece, pp. 1–3, April 2006.
25. T. Sogorb, J.V. Llario, J. Pelegri, R. Lajara, and J. Alberola, "Studying the feasibility of energy harvesting from broadcast RF station for WSN," *IEEE Instrumentation and Measurement Technology Conference*, Canada, pp. 1–3, May 2008.
26. Y. Cao, W. Hong, L. Deng, S. Li, and L. Yin, "A 2.4GHz circular polarization rectenna with harmonic suppression for microwave power transmission," *IEEE International Conference on Internet of Things (iThings) and IEEE Green Computing and Communications (GreenCom) and IEEE Cyber, Physical and Social Computing (CPSCom) and IEEE Smart Data (SmartData)*, China, pp. 1–3, December 2016.
27. T. Mitani, S. Kawashima, and T. Nishimura, "Analysis of voltage doubler behaviour of 2.45-GHz voltage doubler-type rectenna," *IEEE Transactions on Microwave Theory and Techniques*, vol. 65, no. 4, pp. 1051–1057, 2017.
28. Y.S. Chen and C.W. Chiu, "Maximum achievable power conversion efficiency obtained through an optimized rectenna structure for RF energy harvesting," *IEEE Transactions on Antennas and Propagation*, vol. 65, no. 5, pp. 2305–2318, May 2017.
29. V. Kuhn, C. Lahuec, F. Seguin, and C. Person, "A multi-band stacked RF energy harvester with RF-to-DC efficiency up to 84%," *IEEE Transactions on Microwave Theory and Techniques*, vol. 63, no. 5, pp. 1768–1777, 2015.
30. T. Salter, K. Choi, M. Peckerar, G. Metze, and N. Goldsman, "RF energy scavenging system utilizing switched capacitor DC-DC converter," *IET Electronics Letters*, vol. 45, no. 7, pp. 374–376, 2009.
31. F. Kocer and M.P. Flynn, "A new transponder architecture with on-chip ADC for long-range telemetry applications," *IEEE Journal of Solid State Circuits*, vol. 41, pp. 1142–1149, May 2006.
32. G. Andıa Vera, "Efficient rectenna design for ambient microwave energy recycling," *Thesis of Engineering*, Barcelona, Spain, pp. 1–115, July 2009.
33. J.A. Hagerty, F.B. Helmbrecht, W.H. McCalpin, R. Zane, and Z.B. Popovic, "Recycling ambient microwave energy with broad-band rectenna arrays," *IEEE Transactions on Microwave Theory and Techniques*, vol. 52, no. 3, pp. 1014–1024, 2004.
34. M. Zeng, A.S. Andrenko, X. Liu, Z. Li, and H.Z. Tan, "A compact fractal loop rectenna for RF energy harvesting," *IEEE Antennas and Wireless Propagation Letters*, vol. 16, pp. 2424–2427, July 2017.

35. S. Mandal and R. Sarpeshkar, "Low-power CMOS rectifier design for RFID applications," *IEEE Transactions on Circuits and Systems I: Regular Papers*, vol. 54, pp. 1177–1188, June 2007.
36. K. Kotani, A. Sasaki, and T. Ito, "High-efficiency differential drive CMOS rectifier for UHF RFIDs," *IEEE Journal of Solid State Circuits*, vol. 44, pp. 3011–3018, November 2009.
37. V. Marian, B. Allard, C. Vollaire, and J. Verdier, "Strategy for microwave energy harvesting from ambient ELD or a feeding source," *IEEE Transactions on Power Electronics*, vol. 27, pp. 4481–4491, 2012.
38. S. Gaurav, R. Ponnaganti, T.V. Prabhakar, and K.J. Vinoy, "A tuned rectifier for RF energy harvesting from ambient radiations," *AEU-International Journal of Electronics and Communications*, vol. 67, pp. 564–569, 2013.
39. Y.Y. Xiao, Z. X. Du, and X. Y. Zhang, "High-efficiency rectifier with wide input power range based on power recycling," *IEEE Transactions on Circuits and Systems-II: Express Briefs*, vol. 65, pp. 744–748, 2018.
40. HP HSMS2850 Datasheet. http://www.hp.woodshot.com/hprfhelp/4downld/products/diodes/hsms2850.pdf.
41. Skyworks. SMS7630 Datasheet. http://www.skyworksinc.com/uploads/documents/SMS7630061201295H.pdf.
42. I. Chaour, A. Fakhfakh, and O. Kanoun, "Enhanced passive RF-DC converter circuit efficiency for low RF energy harvesting," *Sensors*, vol. 17, no. 546, pp. 1–14, 2017.
43. M.A. Rosli, S.A.Z. Murad, M.N. Norizan, and M.M. Ramli, "Design of RF-to-DC conversion circuit for energy harvesting in CMOS 0.13-µm technology," *AIP Conference Proceedings*, vol. 2045, pp. 020089, 2018.
44. A. Collado, S.N. Daskalakis, K. Niotaki, R. Martinez, F. Bolos, and A. Georgiadis, "Rectifier design challenges for RF wireless power transfer and energy harvesting systems," *Radioengineering*, vol. 26, 1–7, 2017.
45. V. Rizzoli, G. Bichicchi, A. Costanzo, F. Donzelli, and D. Masotti, "CAD of multi-resonator rectenna for micro-power generation," *2009 European Microwave Integrated Circuits Conference* (EuMIC), Rome, Italy, pp. 331–334, 2009.
46. S. Keyrouz, H.J. Visser, and A.G. Tijhuis, "Multi-band simultaneous radio frequency energy harvesting," *7th European Conference on Antennas and Propagation (EuCAP)*, Gothenburg, Sweden, pp. 3058–3061, 2013.
47. R. Scheeler, S. Korhummel, and Z. Popovic, "A dual frequency ultralow-power efficient 0.5-g rectenna," *IEEE Microwave Magazine*, vol. 15, no. 1, pp. 109–114, 2014.
48. T. Oka, T. Ogata, K. Saito, and S. Tanaka, "Triple-band single-diode microwave rectifier using CRLH transmission line," *Proceedings of 2014 Asia-Pacific Microwave Conference*, Sendai, Japan, pp. 1013–1015, 2014.
49. A. Collado and A. Georgiadis, "Conformal hybrid solar and electromagnetic (EM) energy harvesting rectenna," *IEEE Transactions on Circuits and Systems I: Regular Papers*, vol. 60, no. 8, pp. 2225–2234, 2013.
50. D. Belo, A. Georgiadis, and N.B. Carvalho, "Increasing wireless powered systems efficiency by combining WPT and electromagnetic energy harvesting," *IEEE Wireless Power Transfer Conference* (WPTC). Aveiro, Portugal, pp. 3, 2016.
51. Y. Huang, N. Shinohara, and H. Toromura, "A wideband rectenna for 2.4 GHz-band RF energy harvesting," *IEEE Wireless Power Transfer Conference* (WPTC). Aveiro, Portugal, pp. 3, 2016.
52. C.H. Tsai, I.N. Liao, C. Pakasiri, H.C. Pan, and Y.J. Wang, "A wideband 20 mW UHF rectifier in CMOS," *IEEE Microwave and Wireless Components Letters*, vol. 25, no. 6, pp. 388–390, 2015.

53. S. Abbasian and T. Johnson, “High efficiency GaN HEMT synchronous rectifier with an octave bandwidth for wireless power applications,” *IEEE MTT-S International Microwave Symposium* (IMS). San Francisco, CA, pp. 4, 2016.
54. J. Kimionis, A. Collado, M.M. Tentzeris, and A. Georgiadis, “Octave and decade UWB rectifier based on non-uniform transmission lines for energy harvesting,” *IEEE Transactions on Microwave Theory and Techniques*, vol. 65, no. 11, pp. 4326–4334, November 2017.
55. F. Bolos, D. Belo, and A. Georgiadis, “A UHF rectifier with one octave bandwidth based on a non-uniform transmission line,” *IEEE MTT-S International Microwave Symposium* (IMS). San Francisco, CA, pp. 3, May 2016.

4 Rectennas for RFEH Systems

4.1 INTRODUCTION

Fast progress in the wireless devices and systems and the demands of low-power electronic sensors, integrated circuits, and devices have increased drastically. Various research trends have tended to study the feasibility of powering these low-power devices by harvesting ambient/free radio frequency (RF) energy from ambient electromagnetic in environment from various RF sources. Recently, RFEH (RF energy harvesting) technology has received much attention for utilizing clean and renewable power sources. Rectenna (rectifying antenna) system can be used for remotely charging batteries in several sensor networks for Internet of Things (IoT) applications, which are commonly used in smart buildings, implanted medical devices, and automotive applications. Rectenna, which is used to convert RF energy into usable DC electrical energy, is mainly a combination between a receiving antenna and a rectifier circuit. This chapter presents several designs for circularly polarized (CP) rectennas, millimeter-wave rectennas, and wideband, multiband, high-efficiency, and low-power rectennas with different characteristics for RFEH applications.

4.2 CIRCULARLY POLARIZED RECTENNAS

The CP rectennas (Figure 4.1) allow the system to harvest RF energy regardless of the device orientation and different polarized RF waves [1]. As the RF energy found in the surrounding environment can exist in any orientation/polarization and phase alignment or unknown RF sources, CP rectennas are desirable for RFEH systems.

FIGURE 4.1 Compact efficient CP rectenna at 0.9GHz. (From Jie, A.M., Nasimuddin, Karim, M.F., and Chandrasekaran, K.T., *IEEE Antennas Propag. Magazine*, 61, 94–111, 2019. With permission.)

Capability to radiate and receive RF energy in any plane with minimum loss makes CP antennas a key candidate for RFEH system. A compact and efficient CP rectenna (wide-angle CP tapered-slit based antenna with a compact rectifier circuit) at 0.9 GHz UHF band has been investigated for RFEH systems [1], and the rectenna has a high RF-to-DC conversion efficiency of 43%.

A CP beam-switching wireless power transfer system (Figure 4.2) is used for ambient energy harvesting applications operating at 2.4 GHz [2]. A generalized stubs-integrated-microstrip (SIM) antenna with wide coverage CP radiation is proposed for RFEH applications [3]. As museum contents are vulnerable to bad ambience conditions and human vandalization, preserving them is a duty towards humanity. In Ref. [4], authors have developed an IoT-based system for monitoring and controlling the museum's ambient conditions, as shown in Figure 4.3. A novel rectenna using a compact CP patch antenna with RF-to-DC power conversion part at 2.45 GHz is introduced in Ref. [5].

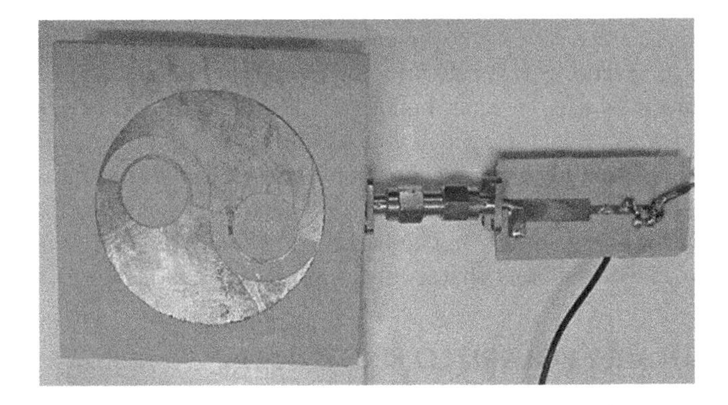

FIGURE 4.2 CP high-gain rectenna at 2.45 GHz. (From Chandrasekaran, K.T., Nasimuddin, Alphones, A., and Karim, M.F., "Compact circularly polarized beam-switching wireless power transfer system for ambient energy harvesting applications," *International Journal of RF and Microwave Computer-Aided Engineering*, January 2019. With permission.)

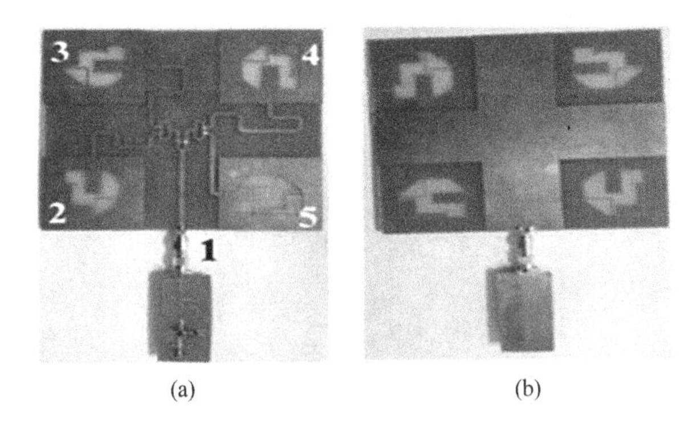

FIGURE 4.3 2×2 array-based CP rectenna (a) top view and (b) bottom view. (From Eltresy, N.A. et al., *Sensors*, 19, 4465, 2019. With permission.)

A dual-band CP rectenna [6] (tapered-slit radiating patch and slotted circular parasitic patch with dual-band rectifier circuit), shown in Figure 4.4, has been proposed at UHF (0.9 GHz) and Wi-Fi band (2.4 GHz). Dual arc-shaped slotted dual-band CP (1.85 and 2.45 GHz) rectenna [7] has been proposed for RFEH (Figure 4.5), a triple-band CP

FIGURE 4.4 Dual-band (900 MHz UHF and 2.45 GHz Wi-Fi) CP rectenna. (From Jie, A.M., Nasimuddin, Karim, M.F., and Chandrasekaran, K.T., *Int. J. RF Microw. Computer-Aided Eng.*, 29, 2019. With permission.)

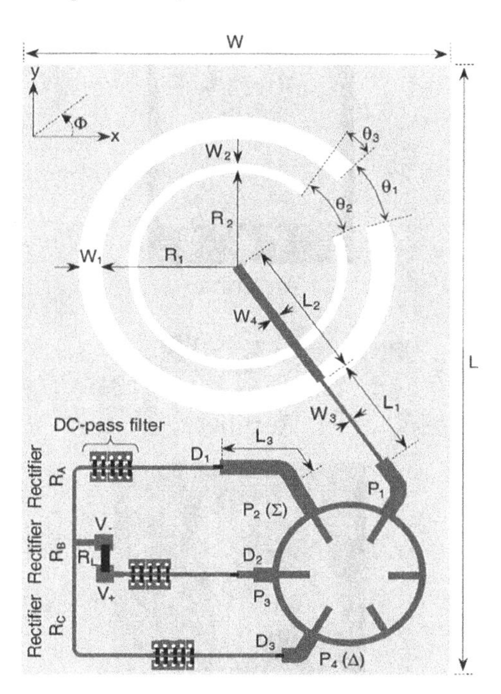

FIGURE 4.5 Dual-band (1.85 and 2.45 GHz). (From Takhedmit, H., Saddi, Z., and Cirio, L., *Prog. Electromagn. Res. C*, 79, 89–100, 2017. With permission.)

rectenna for Wi-Fi (2.4 GHz), Wi-MAX (3.5 GHz), and WLAN (5.2 GHz) (Figure 4.6 [8]), and cross-shaped aperture coupled dual-polarized CP rectenna (Figure 4.7 [9]). In Figure 4.5, rectenna consists of the dual arc-shaped slotted rectenna with dual-band CP radiation and a dual-band RF-to-DC rectifying circuit. It was fabricated by microstrip technology on Arlon 25N substrate (ε_r=3.4, tanδ=0.0025, h=1.524 mm) and operates at 1.85 and 2.45 GHz. The CP is chosen in order to keep an almost constant received RF power level regardless of the orientation of the antenna and the polarization of the incident electromagnetic wave. The antenna is printed on the ground plane, located on the top layer of the circuit. The conversion circuit, fed by a microstrip line, is printed on the bottom layer of the circuit.

A novel six-band dual-CP antenna [10] has been proposed to receive the RF energy from six different frequency bands with almost all polarized waves. Due to the nonlinearity and complex input impedance of the rectifying circuit, the design of a multiband and/or broadband rectenna is always challenging and its performance can be easily affected by variation in the input power level and load. A broadband dual-CP receiving antenna, which has a very wide bandwidth (from 550 MHz to 2.5 GHz) and is compact in size, has been developed.

A novel wideband CP antenna array using sequential rotation feeding network is presented in Figure 4.8 [11] with a relative bandwidth of 38.7% at frequencies from 5.05 to 7.45 GHz with a highest gain of 12 dBi at 6 GHz. In Ref. [12], a metamaterial electromagnetic energy harvester (Figure 4.9) constructed via the capacitive loading of metal circular split rings is presented.

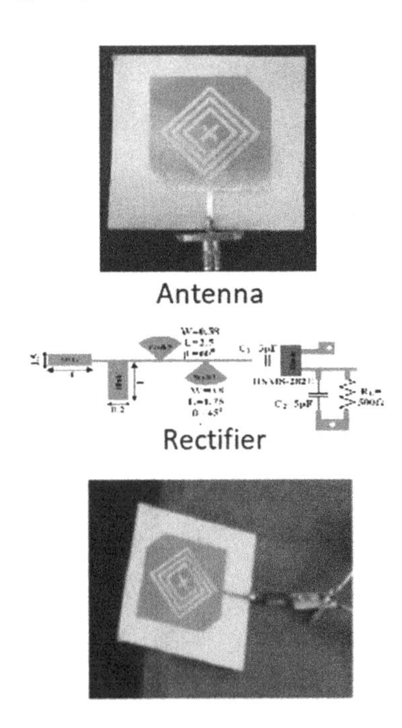

FIGURE 4.6 Tri-band CP rectenna. (From Singh, N., Kanaujia, B.K., Beg, M.T., Mainuddin, and Kumar, S., *Electromagnetics*, 39, 481–490, 2019. With permission.)

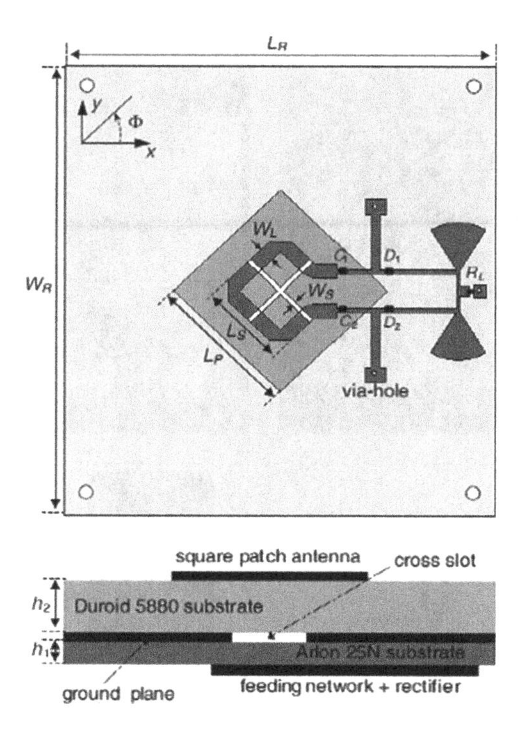

FIGURE 4.7 Dual-polarized CP rectenna. (From Haboubi, W., Takhedmit, H., Luk, J.D.L.S., Adami, S.E., Allard, B., Costa, F., Vollaire, C., Picon, O., and Cirio, L., *Prog. Electromagn. Res.*, 148, 31–39, 2014. With permission.)

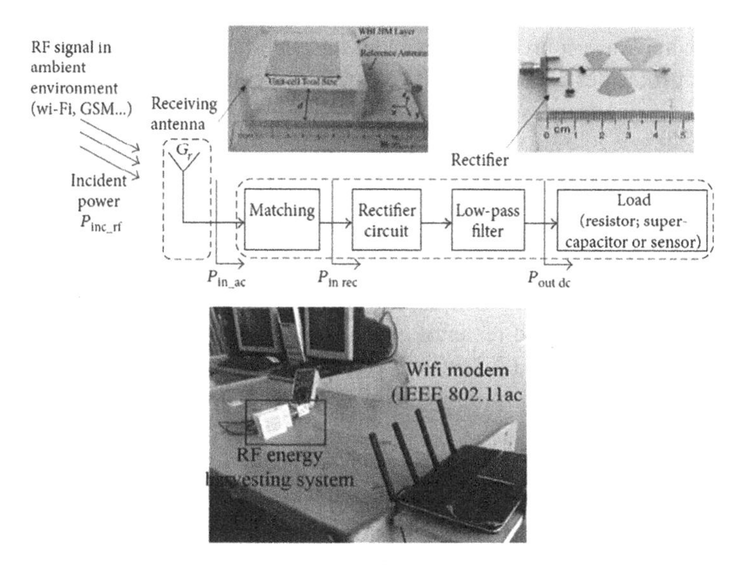

FIGURE 4.8 Wideband high-gain CP antenna for RFEH system. (From Nguyen, N.H., Bui, T.D., Le, A.D., Pham, A.D., Nguyen, T.T., Nguyen, Q.C., and Le, M.T., *Int. J. Antennas Propag.*, 2018, 9, 2018. With permission.)

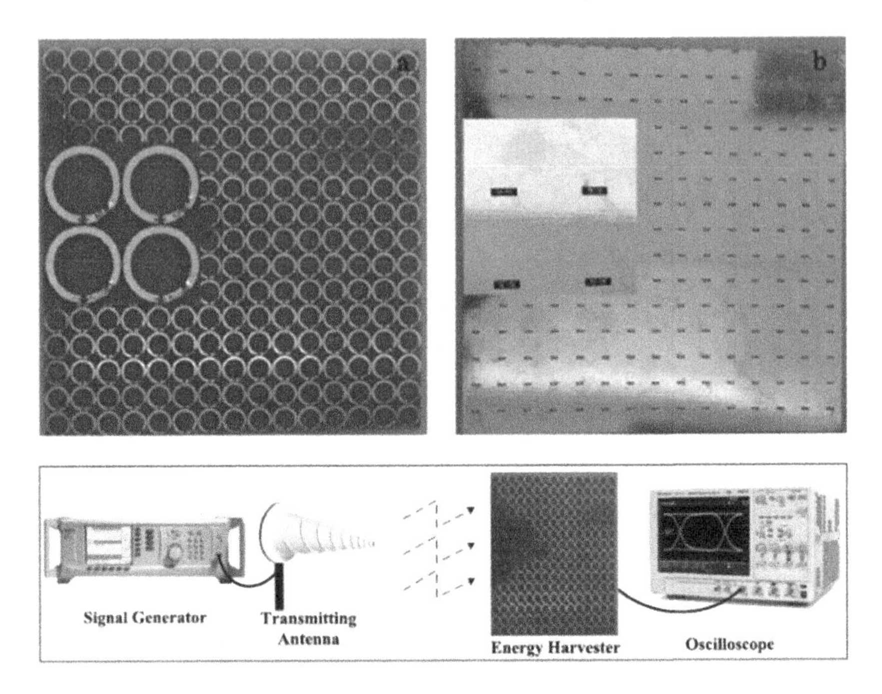

FIGURE 4.9 Photographs of the 15×15 metamaterial harvesting slab (top-layer and bottom-layer) and experimental setup for measuring the voltage induced across the resistive load. (From Shang, S., Yang, S., Liu, J., Shan, M., and Cao, H., *J. Appl. Phys.*, 120, 045106, 2016. With permission.)

Millimeter-wave CP rectenna [13] has been developed at 24 GHz to harvest energy from millimeter-wave sources such as radar and satellite systems. The rectenna circuit structure includes a compact CP substrate integrated waveguide (SIW) based cavity-backed antenna array integrated with a self-biased rectifier using commercial Schottky diodes.

4.3 WIDEBAND RECTENNAS

Wideband rectennas are very useful to collect larger RF energy from wide spectrum (wide frequency range) in the surrounding environment and convert it to DC power. Different types of wideband rectennas have been investigated for RFEH system in the literature. A wideband (2–2.6 GHz) CPW-fed cross-shaped slot antenna based rectennas [14,15] have been proposed for RFEH applications with measured efficiency more than 40%.

A wideband rectenna comprising a wideband cross-dipole antenna, a microwave low-pass filter, and a doubling rectifying circuit using Schottky diodes as rectifying elements has been proposed in Ref. [16] (Figure 4.10). Wideband rectenna design which can harvest the ambient RF power at 1.7–2.5 GHz has been proposed to generate the maximum rectenna conversion efficiency of nearly 57% around 1.7 GHz and over 20% over the wideband.

A novel compact broadband rectenna, shown in Figure 4.11, has been proposed in Ref. [17], and etching fractal geometry-based slot antenna with excellent bandwidth and matching characteristic is used to design the rectenna. A rectifier with a single-stub matching network is investigated for impedance matching, and the RF-to-DC conversion efficiency is improved at low input power.

FIGURE 4.10 CPW-fed wideband cross-shaped slot rectenna. (From Zhang, J.W., Huang, Y., and Cao, P., *Wireless Eng. Technol.*, 5, 107–116, 2014. With permission.)

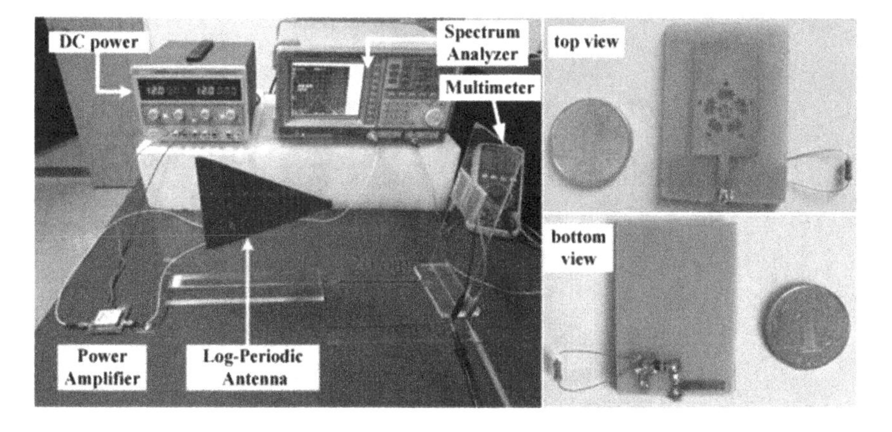

FIGURE 4.11 CPW-fed wideband cross-shaped slot rectenna. (From Shi, Y., Jing, J., Fan, Y., Yang, L., Li, Y., and Wang, M., *AEU Int. J. Electron. Commun.*, 95, 264–270, 2018. With permission.)

Different types of broadband rectennas have been investigated in Ref. [18] for RFEH application to improve the harvesting power. The receiving antenna has a very wide bandwidth (0.5–2.5 GHz) and an average gain of around 3.5 dBi over the frequency band. The performance of the rectenna was directly evaluated in an outdoor ambient environment in Liverpool, UK.

4.4 HIGH-GAIN RECTENNAS

The main function of rectenna technology for RFEH system is to receive high power RF with compact size. An antenna in rectenna is a key component for receiving RF signals. In RFEH systems, a high-gain receiving antenna can enhance the received power, but it requires large antenna aperture or bulky antenna structure. Various high-gain rectennas have been developed based on high-gain receiving antenna for RFEH applications. Recently, the high-gain grid-array antenna (GAA) based rectennas [19,20] have been proposed for RFEH in low power density environment. The GAA without any extra beamforming networks/without phase shifters can be realized beam steering with high gain to capture more RF waves. Benefited from the capability of tilting the beam angle, two isolated ports at two opposite edges of GAA can excite two beams.

A new 2×2 circular microstrip antenna array cube rectenna [21,22] with air dielectric layer for ambient RF energy harvesting has been demonstrated in Figure 4.12. A cube

Measurement setup

FIGURE 4.12 A cube rectenna for RF energy harvesting. (From Zhu, L., Zhang, J., Han, W., Xu, L., and Bai, X., *Int. J. RF Microw. Computer-Aided Eng.*, 29, e21636, 2018. With permission.)

rectenna formed of four antenna and four rectifiers is designed to harvest RF energy, whose maximum output DC voltage is 2.3 V and the maximum output power is 4 mW that can drive four LEDs and an electronic watch.

A 3×2 rectangular patch array with a high gain of 10.3 dBi is used to design high-gain rectenna [23] with three-stage Dickson charge pump circuit for energy harvesting. The rectenna works at 915 MHz. The maximum rectifier efficiency is 41% at input power of 10 dBm. A 35-GHz rectenna using 4×4 patch antenna array has been proposed in Ref. [24], and the maximum RF-to-DC conversion efficiency of 67% with input RF received power of 7 mW can be achieved.

4.5 MULTIBAND RECTENNAS

To improve efficiency of the RFEH system, the receiving multi-band/dual-band antenna-based rectenna in the RFEH system should capture as much as RF energy with different RF sources to improve the output power. Different types of the multiband rectennas, such as dual-band, tri-band, and multi-band, have been developed for RFEH applications.

A multiband dual-polarized rectenna is presented in Figure 4.13 [25] for RF energy harvesting in C-band application range. The receiving antenna of the designed rectenna consists of a truncated corner square patch loaded with several circular slots, L-slots, and U-slots. A proximity-coupled feeding arrangement is used for obtaining a wide impedance bandwidth so that the complete C-band (4–8 GHz) can be covered.

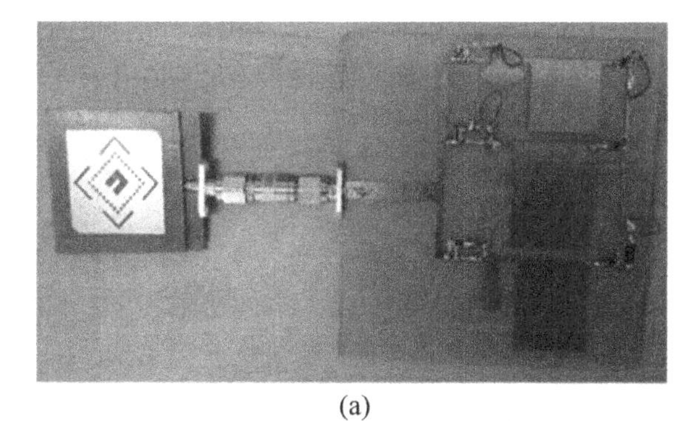

(a)

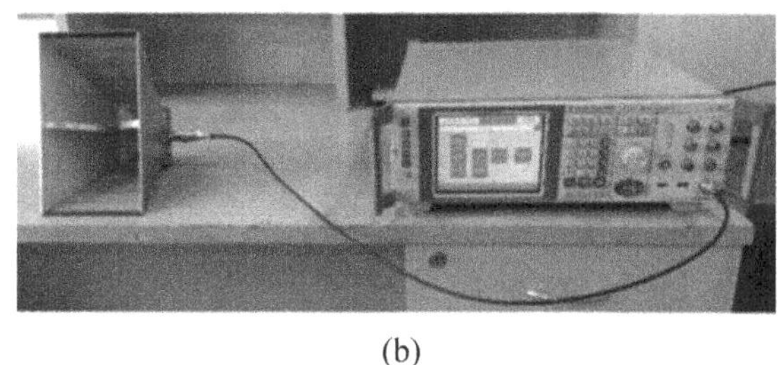

(b)

FIGURE 4.13 Multiband rectenna: (a) Fabricated rectenna and (b) measurement setup. (From Singh, N., Kanaujia, B.K., Beg, M.T., Khan, T., and Kumar, S., *AEU Int. J. Electron. Commun.*, 93, 123–131, 2018. With permission.)

The proposed antenna has the advantage of compact size and dual polarization, since circular polarization is realized at three bands (5.42, 6.9 and 7.61 GHz) in the −10 dB impedance bandwidth range. For efficient RF-to-DC conversion, a two-stage voltage doubler rectifier is used as it provides a higher voltage multiplication with a small threshold voltage at its primary stage. For ensuring maximum RF-to-DC conversion efficiency, a matching network has been designed and connected in between receiving antenna and the rectifier circuitry in order to match the antenna and load in different frequency bands. It is observed that a maximum conversion efficiency of 84% is achieved at 5.76 GHz. The proposed rectenna has been fabricated, and it is found that measured results are in good match with the simulated results. A four-cross-dipole antenna array (4CDAA) based multiband (12, 17, and 20 GHz) rectenna with maximum conversion efficiency of 41% has been proposed for efficient RFEH system in Ref. [26].

A compact dual-band rectenna for GSM900 and GSM1800 energy harvesting has been proposed in Ref. [27], as shown in Figure 4.14. The monopole antenna consists of a longer bent Koch fractal element for GSM900 band and a shorter radiation element for GSM1800. The rectifier is composed of a multi-section dual-band matching network, two rectifying branches, and filter networks. Measured peak efficiency of the proposed rectenna is 62% at 0.88 GHz 15.9 μW/cm^2 and 50% at 1.85 GHz 19.1 μW/cm^2, respectively. Measurement result shows that when the rectenna is 25 m away from a cellular base station, the harvested power is able to power a batteryless LCD watch and achieve 1.275 V output voltage. The proposed rectenna is compact, efficient, low cost, and easy to fabricate, and it is suitable for RF energy harvesting and various wireless communication scenarios.

A monopole-based dual-band (900 and 1800 MHz) rectenna circuit has been proposed in Ref. [28] (Figure 4.15). The rectenna recovers the voltage from 183 to

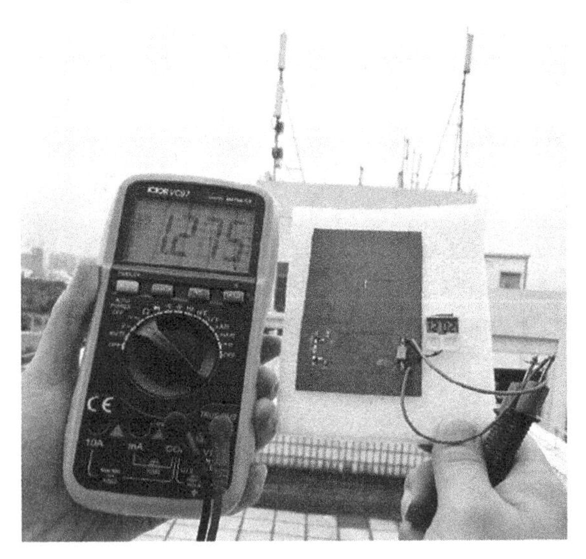

FIGURE 4.14 Fabricated rectenna and measurement setup from cellular base station. (From Zeng, M., Li, Z., Andrenko, A.S., Zeng, Y., and Ta, H.Z., *Int. J. Antennas Propag.*, 2018, 9, 2018. With permission.)

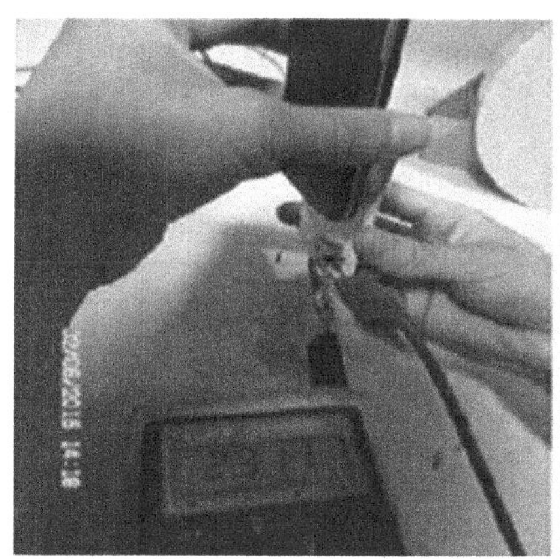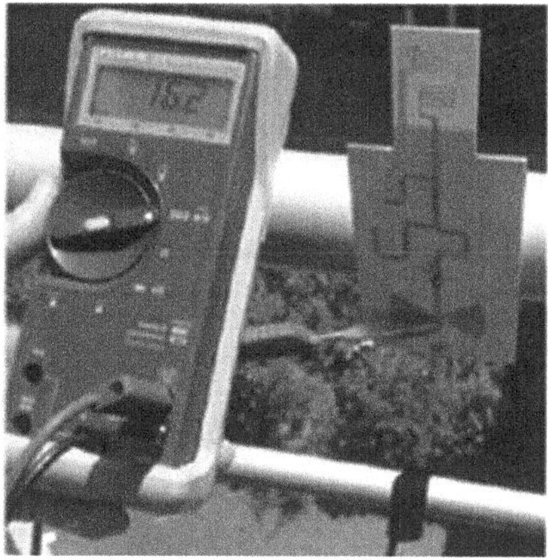

FIGURE 4.15 Fabricated rectenna, rectenna measurement setup in the anechoic chamber at IMEP-LAHC laboratory, and rectenna experiment in ambient environment. (From Ho, D.K., Ngo, V.D., Kharrat, I., Vuong, T.P., Nguyen, Q.C., and Le, M.T., *Adv. Sci. Technol. Eng. Syst. J.*, 2, 612–616, 2017. With permission.)

415 mV when it is placed near the mobile and in ambient environment. A novel compact ultra-lightweight multiband RF energy harvester [29] has been fabricated on a paper substrate. The proposed rectenna is designed to operate in all recently released LTE bands (range 0.79–0.96 GHz; 1.71–2.17 GHz; and 2.5–2.69 GHz). High compactness and ease of integration between antenna and rectifier are achieved by using a topology of nested annular slots.

4.6 HIGH-EFFICIENCY RECTENNAS

A high-efficient broadband rectenna [30] has been proposed for ambient wireless energy harvesting which has a novel broadband rectifying circuit with a new impedance matching circuit designed to match with the ambient RF signals with a relatively low power density. The power sensitivity has been improved by using a full-wave rectifier circuit configuration. A broadband dual-polarization cross-dipole antenna (Figure 4.16) has been designed to enhance the receiving capability of antenna [31]. The harmonic rejection property has been embedded in an integrated antenna by using a novel slot-cutting approach in order to improve the overall efficiency and keep the overall size as small as possible. The simulated and measured results have shown that the rectenna has maximum conversion efficiency of around 55% for –10 dBm input power from 1.8 to 2.5 GHz. The power sensitivity is down to –35 dBm. The rectified DC power can be well above the incident power from any single resource due to the broadband operation and high efficient design. Considering the high DC power output of this design in a relatively low power density environment, this rectenna can be used for efficient wireless energy harvesting for a range of wireless sensor and network applications.

A two-and-a-half-dimensional (2.5D) wafer-level RF energy harvesting rectenna module [32] has been proposed with a compact size and high PCE that integrates a

Rectifying circuit of the proposed rectenna

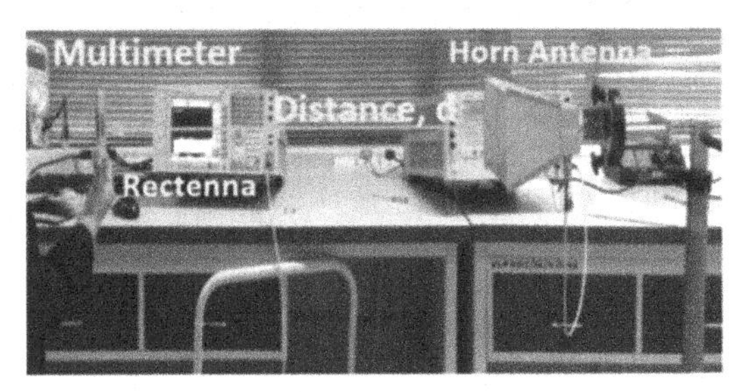

Rectenna measurement setup

FIGURE 4.16 High-efficiency rectenna with experimental setup. (From Meor Said, M.A., *J. Telecommun. Electron. Comp. Eng.*, 9, 151–154, 2017. With permission.)

2.45-GHz antenna in an integrated passive device (IPD) and a rectifier in a tsmcTM 0.18 μm CMOS process. The proposed rectifier provides a master-slave voltage doubling full-wave topology which can reach high PCE by means of a relatively simple circuitry. The IPD antenna is stacked on top of the CMOS rectifier. The rectenna achieves an output voltage of 1.2 V and PCE of 47% when the operation frequency is 2.45 GHz with −12 dBm input power. The peak efficiency of the circuit is 83% with −4 dBm input power. The die size of the RF harvesting module is less than 1 cm^2. The performance of this module makes it possible to power mobile device, and it is also very suitable for wearable and implantable wireless sensor networks (WSN).

For high-efficiency rectennas, various metasurface-based structures with high efficiency [33–36] have been proposed (Figure 4.17) for RFEH applications, in which the distance between two elements is equal to 0.25 mm and each unit cell has been loaded with an 82 Ω resistor. The measured power efficiency of 93% was observed.

4.7 LOW-POWER RECTENNAS

A wearable rectenna to harvest low-power RF energy has been discussed in Ref. [37] for a biomedical device pasted on human body. The wearable textile material,

(a) (b)

FIGURE 4.17 Metasurface-based rectenna with high efficiency (a) top view and (b) bottom view. (From Almoneef, T.S. and Ramahi, O.M., *Appl. Phys. Lett.*, 106, 153902, 2015. With permission.)

cordura fabric, is selected for on-body application. The textile rectenna is designed at the Wi-Fi frequency band, 2.45 GHz, to harvest RF energy. A rectenna (a rectifier with a patch antenna) used here is a linearly polarized. The rectifier is a single-stage full-wave Greinacher rectifier where a radio frequency choke (RFC) is added to increase the output voltage. The output DC voltage of the rectenna can achieve 2.2 V when the RF power is swept from −40 to 0 dBm. The maximum output voltage of 2×2 array rectenna is about 1.05 V as the power transferring distance is 150 cm in the indoor of Wi-Fi environment. An ultra-low power dual-band rectenna has been proposed in Ref. [38].

A flexible 2.45-GHz wireless power harvesting wristband [39] has been proposed to generate a net DC output from a −24.3 dBm RF input. This is the lowest reported system sensitivity for systems comprising a rectenna and impedance-matching power management. A complete system comprising a fabric antenna, a rectifier on rigid substrate, a contactless electrical connection between rigid and flexible subsystems, and power electronics impedance matching has been implemented. Various fabric and flexible materials are electrically characterized at 2.45 GHz using the two-line and the T-resonator methods. Selected materials are used to design an all-textile antenna, which demonstrates a radiation efficiency above 62% on a phantom irrespective of location, and a stable radiation pattern. The rectifier, designed on a rigid substrate, shows a best-in-class efficiency of 33.6% at −20 dBm. A reliable, efficient, and wideband contactless connection between the fabric antenna and the rectifier is created using broadside-coupled microstrip lines, with an insertion loss below 1 dB from 1.8 to over 10 GHz. A self-powered boost converter with a quiescent current of 150 nA matches the rectenna output with a matching efficiency above 95%. The maximum end-to-end efficiency is 28.7% at −7 dBm. The wristband harvester demonstrates net positive energy harvesting from −24.3 dBm, a 7.3 dB improvement on the state of the art.

REFERENCES

1. A.M. Jie, Nasimuddin, M.F. Karim, and K.T. Chandrasekaran, "A wide-angle circularly polarized tapered-slit-patch antenna with compact rectifier for energy harvesting systems," *IEEE Antennas and Propagation Magazine*, vol. 61, no. 2, pp. 94–111, April 2019.
2. K.T. Chandrasekaran, Nasimuddin, A. Alphones, and M.F. Karim, "Compact circularly polarized beam-switching wireless power transfer system for ambient energy harvesting applications," *International Journal of RF and Microwave Computer-Aided Engineering*, vol. 29, pp. e21642, January 2019.
3. S.B. Vignesh, Nasimuddin, and A. Alphones, "Stubs-integrated-microstrip antenna design for wide coverage of circularly polarized radiation," *IET Microwaves, Antennas and Propagation*, vol. 11, pp. 44–49, 2017.
4. N.A. Eltresy, O.M. Dardeer, A. Al-Habal, E. Elhariri, A.H. Hassan, A. Khattab, D.N. Elsheakh, S.A. Taie, H. Mostafa, H.A. Elsadek, and E.A. Abdallah, "RF energy harvesting IoT system for museum ambience control with deep learning," *Sensors (Basel)*, vol. 19, no. 20, pp. 4465, October 2019.
5. T.C. Yo, C.M. Lee, C.M. Hsu, and C.H. Luo, "Compact circularly polarized rectenna with unbalanced circular slots," *IEEE Transactions on Antennas and Propagation*, vol. 56, no. 3, pp. 882–886, March 2008.
6. A.M. Jie, Nasimuddin, M.F. Karim, and K.T. Chandrasekaran, "A dual-band efficient circularly polarized rectenna for RF energy harvesting systems," *International Journal of RF and Microwave Computer-Aided Engineering*, vol. 29, no. 1, January 2019.
7. H. Takhedmit, Z. Saddi, and L. Cirio, "A high-performance circularly-polarized rectenna for wireless energy harvesting at 1.85 and 2.45 GHz frequency bands," *Progress in Electromagnetics Research C*, vol. 79, pp. 89–100, 2017.
8. N. Singh, B.K. Kanaujia, M.T. Beg, Mainuddin, and S. Kumar, "A triple band circularly polarized rectenna for RF energy harvesting," *Electromagnetics*, vol. 39, no. 7, pp. 481–490, 2019.
9. W. Haboubi, H. Takhedmit, J.D.L.S. Luk, S.E. Adami, B. Allard, F. Costa, C. Vollaire, O. Picon, and L. Cirio, "An efficient dual-circularly polarized rectenna for RF energy harvesting in the 2.45 GHz ISM band," *Progress in Electromagnetics Research*, vol. 148, pp. 31–39, 2014.
10. C. Song, Y. Huang, P. Carter, J. Zhou. S. Yuan. Q. Xu, and M. Kod, "A novel six-band dual CP rectenna using improved impedance matching technique for ambient RF energy harvesting," *IEEE Transactions on Antennas Propagation*, vol. 64, no. 7, pp. 3160–3171, July 2016.
11. N.H. Nguyen, T.D. Bui, A.D. Le, A.D. Pham, T.T. Nguyen, Q.C. Nguyen, and M.T. Le, "A novel wideband circularly polarized antenna for RF energy harvesting in wireless sensor nodes," *International Journal of Antennas and Propagation*, vol. 2018, pp. 9, Article ID 1692018, 2018.
12. S. Shang, S. Yang, J. Liu, M. Shan, and H. Cao, "Metamaterial electromagnetic energy harvester with high selective harvesting for left- and right-handed circularly polarized waves," *Journal of Applied Physics*, vol. 120, p. 045106, 2016.
13. S. Ladan, A.B. Guntupalli, and K. Wu, "A high-efficiency 24 GHz rectenna development towards millimeter-wave energy harvesting and wireless power transmission," *IEEE Transaction on Circuits and Systems-I*, vol. 61, no. 12, pp. 3358–3366, December 2014.
14. M.J. Nie, X.X. Yang, G.N. Tan, and B. Han, "A compact 2.45-GHz broadband rectenna using grounded coplanar waveguide," *IEEE Antennas and Wireless Propagation Letters*, vol. 14, pp. 986–989, 2005.
15. N. Saranya and T. Kesavamurthy, "Design and performance analysis of broadband rectenna for an efficient RF energy harvesting application," *International Journal of RF*

and Microwave Computer-Aided Engineering, vol. 29, no. 1, pp. e21628, January 2019, Special Issue: Compact and Efficient RF Energy Harvesting System Designs.
16. J.W. Zhang, Y. Huang, and P. Cao, "An investigation of wideband rectennas for wireless energy harvesting," *Wireless Engineering and Technology*, vol. 5, pp. 107–116, 2014.
17. Y. Shi, J. Jing, Y. Fan, L. Yang, Y. Li, and M. Wang, "A novel compact broadband rectenna for ambient RF energy harvesting," *AEU-International Journal of Electronics and Communications*, vol. 95, pp. 264–270, 2018.
18. C. Song, "Broadband rectifying-antennas for ambient RF energy harvesting and wireless power transfer," Doctor of Philosophy thesis, University of Liverpool, 2017.
19. Y.Y. Hu, S. Sun, H. Xu, and H. Sun, "Grid-array rectenna with wide angle coverage for effectively harvesting RF energy of low power density," *IEEE Transactions on Microwave Theory and Techniques*, vol. 67, no. 1, pp. 402–413, January 2019.
20. Y.Y. Hu, H. Xu, H. Sun, and S. Sun, "A high-gain rectenna based on grid-array antenna for RF power harvesting applications," *10th Global Symposium on Millimeter-Waves*, Hong Kong, China, pp. 1–3, 24–26 May, 2017.
21. L. Zhu, J. Zhang, W. Han, L. Xu, and X. Bai, "A novel RF energy harvesting cube based on air dielectric antenna arrays," *International Journal of RF and Microwave Computer-Aided Engineering*, vol. 29, pp. e21636, September 2018.
22. F. Xie, G. Yang, and W. Geyi, "Optimal design of an antenna array for energy harvesting," *IEEE Antennas and Wireless Propagation Letters*, vol. 12, pp. 155–158, 2013.
23. A. Mavaddat, S.H.M. Armaki, and A.R. Erfanian, "Millimeter-wave energy harvesting using 4×4 microstrip patch antenna array," *IEEE Antennas and Wireless Propagation Letters*, vol. 14, pp. 515–518, 2015.
24. N. Singh, B.K. Kanaujia, M.T. Beg, T. Khan, and S. Kumar, "A dual polarized multiband rectenna for RF energy harvesting," *AEU-International Journal of Electronics and Communications*, vol. 93, pp. 123–131, September 2018.
25. A. Okba, A. Takacs, H. Aubert, S. Charlot, and P.F. Calmon, "Multiband rectenna for microwave applications," *Energy and Radiosciences*, Vol. 18, Issue 2, pp, 107–117, February 2017.
26. M. Zeng, Z. Li, A.S. Andrenko, Y. Zeng, and H.Z. Ta, "A compact dual-band rectenna for GSM900 and GSM1800 energy harvesting," *International Journal of Antennas and Propagation*, vol. 2018, pp. 9, 2018, Article ID 4781465.
27. D.K. Ho, V.D. Ngo, I. Kharrat, T.P. Vuong, Q.C. Nguyen, and M.T. Le, "A novel dual-band rectenna for ambient RF energy harvesting at GSM 900MHz and 1800MHz," *Advances in Science, Technology and Engineering Systems Journal*, vol. 2, no. 3, pp. 612–616, 2017.
28. V. Palazzi, J. Hester, J. Bito, F. Alimenti, C. Kalialakis, A. Collado, P. Mezzanotte, A. Georgiadis, L. Roselli, and M.M. Tentzeris, "A novel ultra-lightweight multiband rectenna on paper for RF energy harvesting in the next generation LTE bands," *IEEE Transactions on Microwave Theory and Techniques*, vol. 66, no. 1, pp. 366–379, January 2018.
29. C. Song, Y. Huang, J. Zhou, J. Zhang, S. Yuan, and P. Carter, "A high-efficiency broadband rectenna for ambient wireless energy harvesting," *IEEE Transactions on Antennas and Propagation*, vol. 63, pp. 3486–3495, 2015.
30. M.A. Meor Said, "A high-efficiency rectenna design at 2.45GHz for RF energy scavenging," *Journal of Telecommunication, Electronic and Computer Engineering*, vol. 9, pp. 151–154, 2017.
31. K.C. Lin, P.C. Wu, H.H. Tsai, Y.Z. Juang, C.S. Yang, and G.W. Huang, "A compact size and high efficiency CMOS-IPD rectenna using 2.5D wafer-level packing for a wireless power harvesting system," *IEEE 46th European Microwave Conference* (EuMC), London, UK, pp. 1–4, 2016.

32. T.S. Almoneef and O.M. Ramahi, "Metamaterial electromagnetic energy harvester with near unity efficiency," *Applied Physics Letters*, vol. 106, no. 15, pp. 153902, 2015.
33. B. Ghaderi, V. Nayyeri, M. Soleimani, and O.M. Ramahi, "A novel symmetric ELC resonator for polarization-independent and highly efficient electromagnetic energy harvesting," *IEEE MTT-S International Microwave Workshop Series on Advanced Materials and Processes for RF and THz Applications* (IMWS-AMP 2017), Pavia, Italy, pp. 20–22, 2017.
34. M. El Badawe and O.M. Ramahi, "Efficient metasurface rectenna for electromagnetic wireless power transfer and energy harvesting," *Progress in Electromagnetics Research*, vol. 161, pp. 35–40, January 2018.
35. G.T. Oumbe Tekam, V. Ginis, J. Danckaert, and P. Tassin, "Designing an efficient rectifying cut-wire metasurface for electromagnetic energy harvesting," *Applied Physics Letters*, vol. 110, no. 8, pp. 1–5, 2017.
36. C.H. Lin, C.W. Chiu, and J.Y. Gong, "A wearable rectenna to harvest low-power RF energy for wireless healthcare applications," *IEEE 11th International Congress on Image and Signal Processing, Bio Medical Engineering and Informatics* (CISP-BMEI), Beijing, China, pp. 1–3, 2018.
37. J. Sampe, N.H.M. Yunus, J. Yunas, and A. Pawi, "Ultra-low power RF energy harvesting of 1.9GHz & 2.45GHz narrow-band rectenna for battery-less remote control," *International Journal of Information and Electronics Engineering*, vol. 7, no. 3, pp. 118–122, May 2017.
38. S.E. Adami, P. Proynov, G.S. Hilton, G. Yang, C. Zhang, and D. Zhu, "A flexible 2.45-GHz power harvesting wristband with net system output from –24.3 dBm of RF power," *IEEE Transactions on Microwave Theory and Techniques*, vol. 66, no. 1, pp. 380–395, January 2018.

Part B

Elements of Wireless Power Transfer (WPT) Systems

5 Antennas for WPT Systems

5.1 INTRODUCTION

Wireless power transmission (WPT) is the technique to transfer power for short or long range without using conductor from transmitter to receiver. It gives freedom from cables and is popular for wireless charging and batteryless implantable devices [1]. The WPT systems are classified into two categories: far-field and near-field systems. Based on the technology, power can be transferred wirelessly using resonant inductive coupling, magnetic resonance coupling, and microwave power transfer [2]. An inductively coupled WPT system has a pair of coupled coils. A magnetic resonant coupling WPT system uses a pair of coupled coils with additional capacitance, which makes the transmitter and the receiver to have the same resonant frequency [3]. Antennas play a great role in WPT systems at both transmitting and receiving ends to transmit and receive signals. Various types of antennas have been developed with the aim of wirelessly transmitting and receiving signals and power. Generally, for a coupled resonant WPT system, antennas of similar type are preferred for transmitting and receiving purposes. Magnetically coupled resonators have shown the capability to transfer power over a longer distance than those of inductive coupling, with higher efficiency than those of RF radiation approach [4].

5.2 ANTENNAS FOR NEAR-FIELD WPT

Near-field WPT systems, also known as non-radiative coupling-based systems, are divided into inductively coupled and magnetically coupled resonant WPT systems [5]. Non-resonant induction can only operate within a close distance (less than the device dimension), whereas resonance coupling can be utilized to considerably increase the distance to mid-range, a factor of at least twice to thrice larger than the device dimension, as the transmitter and the receiver work at same resonant frequency [6,7]. The near-field power is attenuated according to the cube of the reciprocal of the charging distance. In this case, generally coil-based antennas are utilized for power transfer. Helix-type loop antenna [8], Archimedean spiral antenna [9], helical antenna [10], and tape-wound spiral antenna [11] have been utilized for non-radiative power transfer.

Based on loop antenna theory, Fotopoulou and Flynn [12] presented an analytical model for near-field magnetic coupling incorporating misalignment of the RF coil system. They also derived formulae for the magnetic field at the receiver coil when it is laterally and angularly misaligned from the transmitter. They proposed that the presented novel analytical model for near-field magnetic coupling, incorporating misalignment effects, can be used for wireless powering of RFID and implanted biomedical sensors. The efficiency optimization of multiple antennas by transmitting circuit considering the effects of cross-coupling and the sign of mutual inductances was proposed by Imura and Hori [13]. They used the equivalent circuits of multiple antennas to calculate the parameters of the impedance matching circuit, which were used to optimize the transmitting antenna to achieve high efficiency. Antenna geometry for coupled resonant wireless power transfer was discussed by Hirayama et al. [14]. They compared open-ended self-resonant helical antennas and short-ended capacitor-loaded helical antennas from the viewpoint of undesired emission, effect of conductivity, transfer distance, and effect of human body.

The equivalent circuit of a repeater antenna was proposed by Imura [15], including the sign of the mutual inductance, which occurs when repeater antennas are used and is decided by the position of the antenna. Repeater antennas have been proposed to extend the length of the air gap. Long-distance power transfer is achieved by simply installing a repeater between the transmitting and receiving stations. Sample et al. [6] proposed a new analysis method that yields critical insight into the design of practical systems, including the introduction of key figures of merit that can be used to compare systems with vastly different geometries and operating conditions to achieve a near-constant efficiency of over 70% for a range of 0–70 cm. An efficient WPT system was proposed and verified by Park et al. [8], for which it was deduced that a class-D power amplifier (PA) has an advantage as a source when the input resistance changes with the position of the receiving antenna. They investigated a method used to achieve efficient wireless power transfer over a near-field region when the distance between the antennas varies. First, an analysis of the characteristics of the two coupled antennas was performed. Then, the conditions that achieve an efficient WPT system for the load resistance and mutual coupling between the antennas were suggested. Then, they compared several types of PAs as a source of the WPT system. It was shown that the class-D PAs have an advantage in regard to the efficiency for a varied load resistance. A WPT system composed of an unbalanced-fed, ultra-low-profile inverted-L antenna on a rectangular conducting plane was proposed and analyzed numerically by Taguchi and Hirata [16], where the input impedances of the transmitting and receiving antennas were matched to 50 ohms by adjusting the length of the horizontal element and the feed point position. An efficiency of 92.2% was achieved at the design frequency of 100 MHz. A chip-to-package wireless power transfer concept applied to MMIC and antennas on an LCP substrate was presented by Aluigi et al. [9], in which an Archimedean spiral antenna matched to a heterogeneous transformer, which couples the power received by the antenna to the chip, was simulated at the working frequency of 35.4 GHz. Lee et al. [17] presented a system that supplies power to a rotating spindle's diagnostic sensor using wireless power transfer technology based on electromagnetic coupled resonance and gauges the strain experienced by the spindle by attaching power-receiving coils to the spindle at 120° intervals and fixing the power-transmitting coils to the ground at 60° intervals.

Open-end and short-end helical antennas for coupled resonant wireless power transfer were discussed by Hirayama et al. [10] from the viewpoint of undesired emission and biological effect, as shown in Figure 5.1. Due to the effects of human body, the resonant frequency is varied for the open-end model, whereas the loss power is

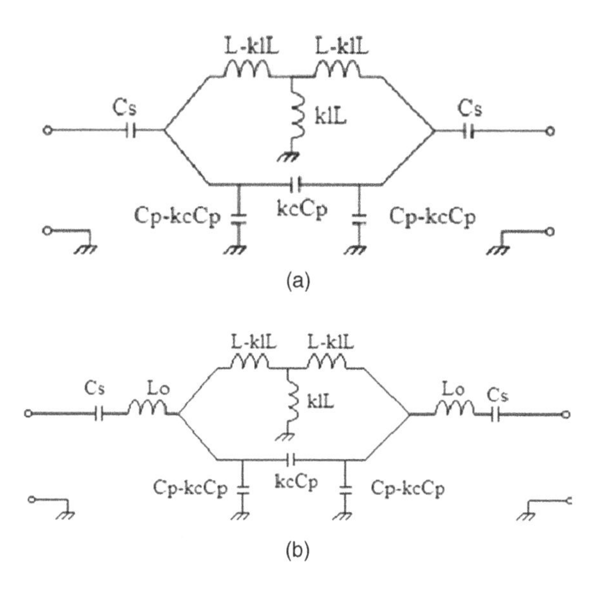

FIGURE 5.1 Equivalent circuit considering E-field and H-field coupling: (a) open-end model; (b) short-end model. (From Hirayama, H., Amano, T., Kikuma, N., and Sakakibara, K., "Undesired emission and biological effect of open-end and short-end antennas for coupled-resonant wireless power transfer," *2013 Asia-Pacific Symposium on Electromagnetic Compatibility* (APEMC), 2013. With permission.)

increased for the short-end model. Phokhaphan et al. [18] proposed a wireless power transfer system (Figure 5.2) which used a printed circuit board as an antenna for both the transmitter and receiver. The antenna in the proposed system was driven by a high-frequency inverter that operates at the resonant frequency of the antenna.

In Ref. [19], the authors proposed a new design approach that uses anti-parallel resonant loops for contactless energy transfer (CET). The forward and reverse loops forming an anti-parallel resonant structure stabilized the transfer efficiency and therefore prevented it from dramatic distance-related changes, a phenomenon that can occur in CET systems with non-radiative methods (or resonant methods). The anti-parallel resonant loops provided an improved efficiency of 87% and six times the frequency insensitivity compared to unidirectional resonant loops within the transferring range without the need for additional automatic resonant circuits. An inductive coupling WPT technology was presented by Mayordomo et al. [20] to allow structural health monitoring of fiber composite structures so that it can be embedded into the structure without compromising its mechanical properties (Figure 5.3).

Chung et al. [21] presented the feasibility of technical fusion between WPT system and superconducting technology, which can be applied to electric charging systems based on resonance method of contactless power transfer technique since it makes possible a convenient charging system. From this point of view, they proposed the combination of WPT technology and high-temperature superconducting (HTS) transmitting antenna, and it is called superconducting WPT system.

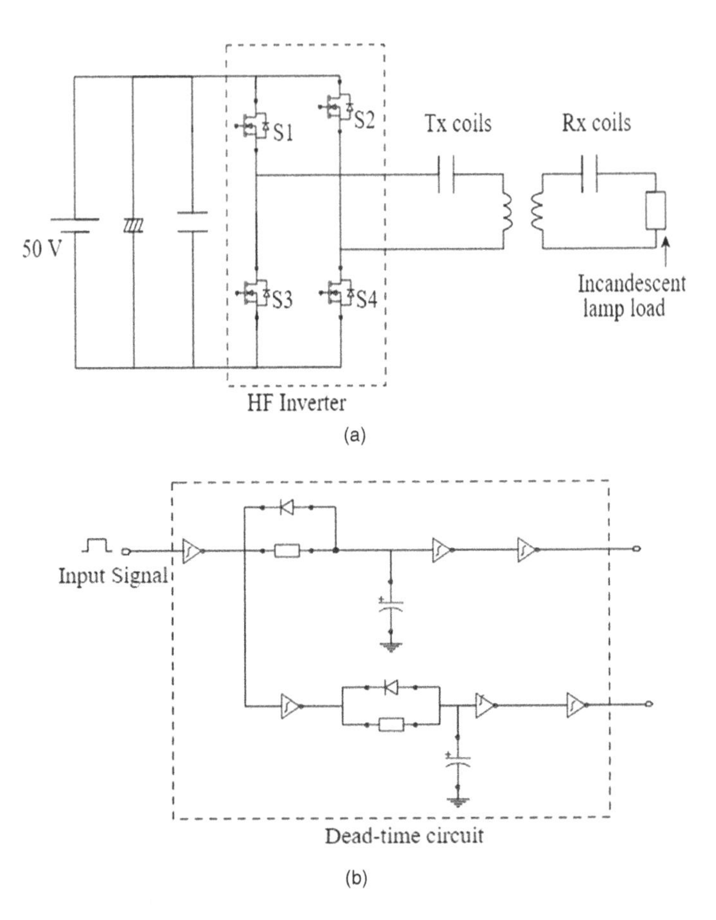

FIGURE 5.2 Configuration of prototype circuit: (a) main circuit; (b) control circuit. (From Phokhaphan, N., Choeisai, K., Noguchi, K., Araki, T., Kusaka, K., Orikawa, K., and Itoh, L.I., "Wireless power transfer based on MHz inverter through PCB antenna," *2013 1st International Future Energy Electronics Conference* (IFEEC), 2013. With permission.)

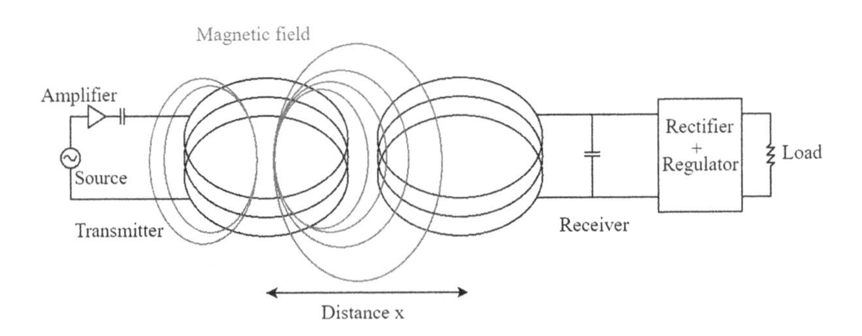

FIGURE 5.3 Inductively coupled WPT system. (From Mayordomo, I., Drager, T., Alayon, J.A., and Bernhard, J., "Wireless power transfer for sensors and systems embedded in fiber composites," *2013 IEEE Wireless Power Transfer* (WPT), 2013. With permission.)

This new approach enables transferring large power efficiently with HTS resonance antennas. EMC characteristics of CRWPT antennas were investigated by Hirayama et al. [11]. They discussed the difference between the self-resonant antenna and the capacitor-loaded inductor by using coupling and resonance model from the viewpoint of EMC. In Ref. [22], a new approach to designing a wireless power transfer system using helical coil (transmitter) and LC resonant coil (receiver) was proposed, as shown in Figure 5.4. The proposed system was optimized at the self-resonant frequency of 3.4 MHz. The helical resonators for wireless power transfer have great benefits of low resonant frequency and simple configuration. It was also shown that the transmit coil with a helical resonator can easily be expanded to multi-channel wireless power transmitters.

Choi et al. [23] proposed a coil-based loop antenna for magnetically coupled WPT systems and predicted the power transfer efficiency at a specific target distance by the calculation of mutual inductance using Neumann's formula and the equivalent model of magnetically coupled WPT system. A new configuration of compact planar coil antennas was proposed by Pu et al. [24] for the design of WPT systems. Based on this, two cases of resonant mid-range WPT systems suitable for small and large load resistors can both be realized efficiently. It was shown that the maximum transfer efficiencies of the proposed WPT systems with loads of 320 m Ω and 50 Ω arrive at around 49.1% and 56.9%, respectively, with the resonant frequency of 19.22 MHz. A method was presented by Kim et al. [25] to analyze the maximum power transfer efficiency of a wireless power transfer system and its electromagnetic fields via spherical modes, and the Z-parameter and Y-parameter for two coupled antennas were derived using the antenna scattering matrix and an addition theorem (Figures 5.5 and 5.6).

A A single antenna based WPT system was proposed by Keskin and Liu [5] to tune the system to its resonant frequency of 13.56 MHz achieved by using radio waves, microwave radiation as well as by applying resonant or inductive coupling. It was successfully demonstrated that the proposed structure can provide the compensation when the system is out of tune from the desired resonant frequency. These structures

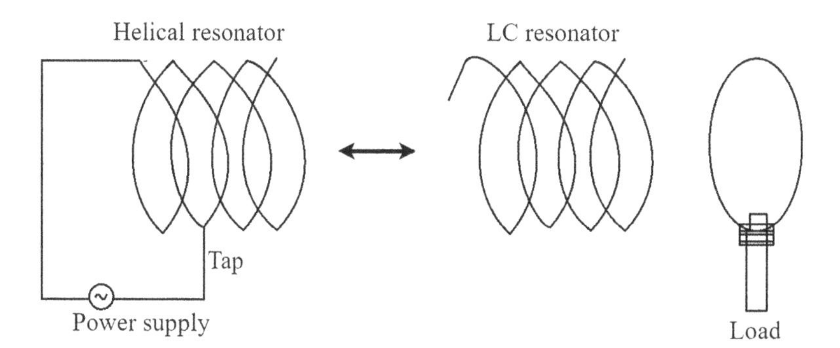

FIGURE 5.4 Schematics of the WPT system using helical resonator and LC resonator. (From Kim, Y., Lee, H., Bang, J., and Chung, C., "A new design of wireless power transfer system using hellical resonators applicable to multi-channel power transmission," *2014 IEEE Wireless Power Transfer Conference*, Jeju, 277–279, 2014. With permission.)

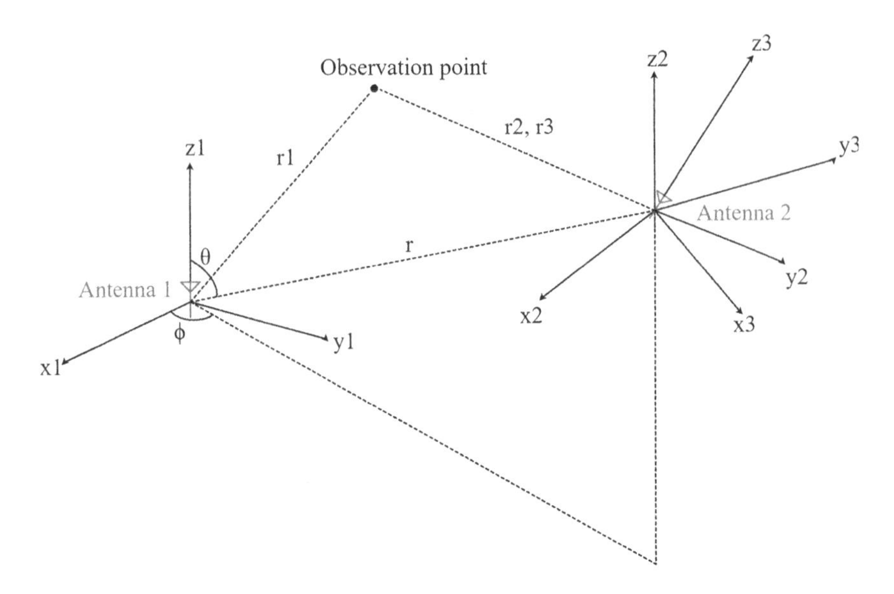

FIGURE 5.5 Coordinate systems and antennas. (From Kim. Y.G. and Nam, S., *IEEE Trans. Antennas Propag.*, 62, 3054–3063, 2014. With permission.)

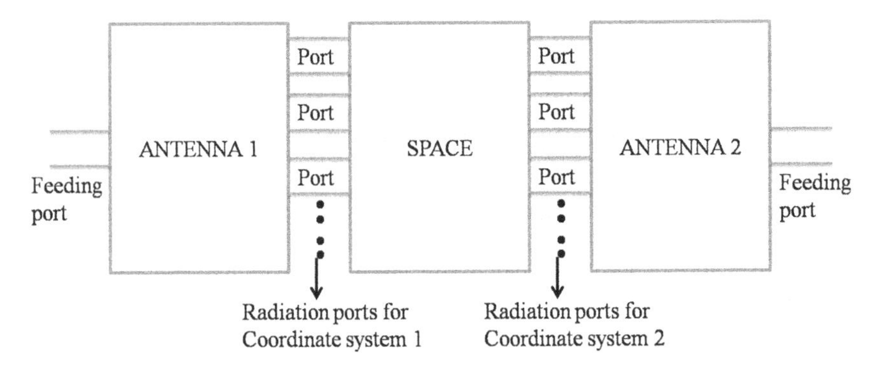

FIGURE 5.6 Network representation of two coupled antennas. (From Kim. Y.G. and Nam, S., *IEEE Trans. Antennas Propag.*, 62, 3054–3063, 2014. With permission.)

enable the designers to optimize both transmitters and receivers by connecting/disconnecting a certain number of unit antenna elements. The proposed method can be used for multi-frequency tuning of different receivers. A WPT system for a pacemaker (PM) was presented by Das and Yoo [26] by employing a resonant inductive coupling method in the WPT system by introducing a spiral transmitter (Tx) coil and a spiral receiver (Rx) coil. In order to increase the efficiency of the WPT, the concept of Yagi–Uda antenna using metamaterials (MTMs) was introduced. A coupled resonant wireless power transfer (WPT) system using a shielded loop antenna for high-efficiency transmission was proposed by Kajiura and Hirayama [27]. The transmission efficiency of the system was determined by the coupling coefficient k and the Q-factor (Figure 5.7).

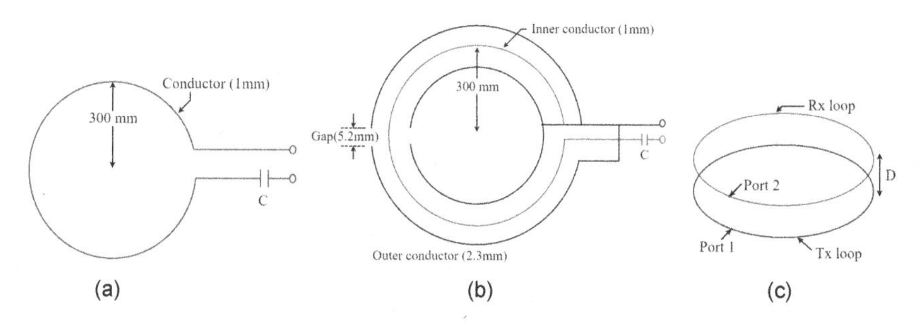

FIGURE 5.7 (a) Conventional loop antenna; (b) shielded loop antenna; (c) simulated model. (From Kajiura, N., and Hirayama, H., "Improvement of transmission efficiency using shielded-loop antenna for wireless power transfer," *Proceedings of ISAP2016*, Okinawa, Japan, 2016. With permission.)

Ando et al. [28] proposed a folded spiral antenna based on the self-resonant spiral antenna that has folded structure with conductors of different diameters for WPT. By using the proposed structure, a maximum efficiency was achieved without using a matching circuit. Numerical simulations demonstrated that by using the proposed structure, the efficiency can be maintained above 80% up to a 3.3 times increase in transfer distance. Nakamura et al. [29] proposed a tape-wound dielectric-loaded spiral antenna for coupled resonant WPT to realize low-frequency high-efficiency power transmission. This antenna was fabricated by lap winding a copper tape and a dielectric tape. Because of the tape structure and the dielectric loading effect, the self-resonant frequency decreased by 30.8%, which enables us to decrease the loss in an RF power source.

An innovative design of a spiral electrically small magnetic antenna (ESMA) with high radiation efficiency was proposed for WPT by Huang and Li [30], which consisted of an embedded multi-loop feed and an innovative spiral resonator. There are two innovation points for the developed ESMA. The first innovation point is the technique of increasing radiation efficiency: (i) Instead of wires, conductor strips are used to reduce loss. (ii) The internal and external turns stagger in the axial direction, and the electric current in the internal turn is not shielded by the external one. The inductance of the feed loop is increased by using embedded multi-loop structure, and the mutual inductance between the feed loop and the resonator is added, then the input impedance is made closer to 50 Ω. A symmetric dipole antenna pair employing kQ product as an alternative index to maximum power coupling efficiency was presented by Ohira and Sakai [31], and electromagnetic field simulation was carried out to deduce the antenna's self- and mutual impedances.

Heo et al. [32] presented a wearable textile antenna embroidered on a fabric for wireless power transfer systems, and a planar spiral coil was generated with the conductive thread on a cotton substrate and was connected to a rectifier circuit fabricated on a flexible polyethylene terephthalate film to constitute a bendable receiver by the magnetic resonance. In Ref. [33], four newly designed antenna configurations for the applications of short-distance WPT were proposed. The antennas were developed from a traditional circular loop antenna, and all of them had a structural feature in common; that is, all of them were in enclosed shape. Compared with the conventional loop antenna, the four always performed better.

Lu et al. [34] proposed a resonant modeling of WPT with a miniaturized antenna using I-type ferrite core in radio propagation for ISM band. It uses mutual inductance between two ferrite core antennas to increase the transmission efficiency. In Ref. [7], a switched-capacitor-based stimulator circuit that enables efficient energy harvesting for neuro-stimulation applications was presented, followed by the discussion on the optimization of the inductive coupling front-end through a co-design approach. The stimulator salvages input energy and stores it in a storage capacitor and, when the voltage reaches a threshold, releases the energy as an output stimulus. The dynamics of the circuit are automatically enabled by a positive feedback, eliminating any stimulation control circuit blocks.

In Ref. [35], an antenna alignment method (Figure 5.8) using intermodulation was proposed to address WPT antenna misalignments in a safe way without increasing the incident power density. Two-tone (2T) waveform excitation was employed to enhance the rectification as well as to generate intermodulation. The intermodulation power was fed back via a magnetic resonant coupling link. Due to the monotonic relation between the intermodulation power and the degree of antenna misalignments, the receiving antenna (in body) can be aligned with the transmitting one (out body). Such a scheme has the advantages of less harm to human tissue and less interference between the 2T waveform excitation and the magnetic resonant coupling link. In Ref. [36], a new approach for the performance improvement of WPT systems by utilizing the high permittivity property of metamaterials (MTMs) was presented. They considered fully planar structures to design the transmitting (Tx) and receiving (Rx) sections, and the performance was enhanced by placing the MTM slabs at both sides of the Tx and Rx elements. The high permittivity property of MTMs generates a strong magnetic dipole behavior, which increases the magnetic coupling between the Tx and Rx sections. As a result, the efficiency of the proposed system was improved significantly along with the transmission distance. Lin et al. [37] designed an antenna pair for short-distance WPT in the reactive near-field (NF) region. Instead of designing each antenna alone to achieve good reflection coefficients, the transmitting (Tx) and receiving (Rx) antennas as a pair are designed to maximize the transmission coefficients. Thus, the antenna pair can mutually compensate for the impedance mismatching caused by the coupling between them to enhance the

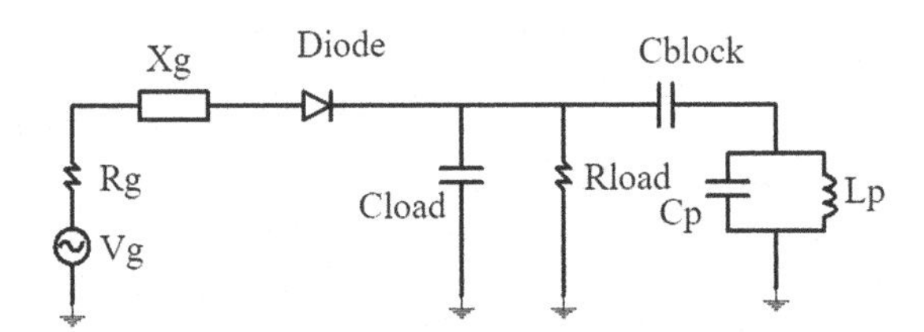

FIGURE 5.8 Small-signal equivalent circuit for single series-mounted rectifier under 2T waveform excitation. (From Zhang, H., Gao, S., Ngo, T., Wu, W., and Guo, Y., *IEEE Trans. Micro. Theory Tech.*, 67, 1708–1716, 2019. With permission.)

energy transfer efficiency. Based on this concept, the developed Rx antenna, a key component of rectennas, has a dual functionality to retain relatively stable reflection coefficients when it is moved far away from the NF Tx antenna.

5.3 ANTENNAS FOR RADIATIVE FAR-FIELD WPT

The far-field WPT has rapidly been developed as it provides longer operational distance. Single-band antennas, multi-band antennas, circularly polarized antennas, and array antennas are used for this type of WPT system.

The power transmitting and receiving coils should be positioned at the same direction in the WPT system using electromagnetic induction; otherwise, losses will occur [38]. Also, it is applicable for short-range applications. The far-field WPT has rapidly been developed as it provides longer operational distance [39]. Antennas are an integral part of the far-field WPT systems, and also the complexity of the system depends on the antenna structure. Single-band antennas, multi-band antennas, circularly polarized antennas, and antenna arrays are used for this type of WPT system.

5.3.1 Multi-Frequency Antennas

As antennas are the critical components of the WPT systems, their operating frequency, gain, radiation pattern, and polarization characteristics highly affect the system performance. In fact, antennas operating at lower frequencies are preferred because of their safety to the human body and transmission characteristics [38]. Various single-band antennas operating at different frequencies have been reported for data telemetry [39], powering low-power wireless sensors [40], biomedical applications [41], etc. The WPT system must be efficient, and the loss of energy during the whole process should be minimized by suppressing the undesired signal. However, the rectenna system has a harmonic suppression filter that contributes additional insertion loss and requires extra area. Hence, the antenna having harmonic suppression characteristics is used to avoid the need of the extra filter component [42].

Kumagai et al. [38] examined a wireless power transmission to a capsular endoscope by electromagnetic waves to show its usability for medical applications and proposed a modified helical antenna inside the endoscope as a power-receiving antenna, operating at 915 MHz.

A reconfigurable stacked patch antenna operating at 5.8 GHz with 9.4 dBi gain and 7.6% bandwidth was introduced by Yang et al. [43] for wireless power reception and data telemetry applications in sensors. At a lower frequency of 2.45 GHz, the antenna operated as a planar inverted-F antenna (PIFA) with 3.3 dBi gain and 2.0% bandwidth. Switching between the two regimes of operation was achieved using PIN diodes. It was proposed that the antenna can be used for wireless power reception in sensors at 5.8 GHz and for data telemetry in between a sensor and a control station at 2.45 GHz. Popovic et al. [40] presented an overview of the far-field wireless powering for low-power wireless sensors with expected RF power densities in the 20–200 μW/cm^2 range in which the sensor platform was powered through an antenna.

The antenna received incident EM waves in the GHz frequency range and coupled the energy to a rectifier circuit which charged a storage device (e.g., thin-film battery) through an efficient power management circuit, and the entire platform was controlled through a low-power microcontroller. Takeiet al. [41] proposed a receiving antenna that operated in 430 MHz and stowed it in the capsular endoscope to transceive wireless signal for medical applications. Moreover, a maximum received power of 35 mW can be reached to power on the capsular endoscope, when the input power of the transmitting antenna is 1.6 W. Therefore, the proposed receiving antenna is suitable for WPT to the capsular endoscope. Zainol et al. [42] presented antenna harmonic suppression for wireless power transfer systems and demonstrated the application of open stub on the feed line, curvature slots, and notch at the antenna patch suppressed undesired signals exhibit near second order effectively.

An RFID reader antenna with a truncated segmented coupling elliptical loop used to achieve inductive WPT was proposed by Wang et al. [44] to operate at UHF band. The WPT feature during communication makes the proposed antenna desirable for some special RFID applications such as implantable medical devices without battery. A 94-GHz microstrip rectenna was developed by MEMS technology for a high-power WPT system by Matsui et al. [45]. To reduce the fabrication cost, a fin-line was also developed for RF power transmission from the waveguide to the microstrip line. Dias et al. [46] proposed a lens antenna for WPT operating at 20 GHz using 3D printing technology. Lens is used to focus the beam of a lower-gain feed antenna to produce a highly directive pattern with low side lobe. Goncalves et al. [47] presented the design of an antenna, operating in the Ku-band, conceived for wireless power transfer systems. The proposed antenna comprised of a hemispherical dielectric lens, fed by a microstrip patch antenna array fabricated using 3D printing technology. The dielectric lens can be coupled with slot-fed microstrip patches to enhance the radiation gain and therefore to increase the efficiency of the wireless power transfer system. Belamgi et al. [48] proposed a suspended planar patch antenna for ISM band (2.4–2.485 GHz) applications, in which the radiating patch of the suspended patch antenna was placed over a ground plane in air at a height and was fed by a coaxial probe. Kumar et al. [49] presented a ring-loaded arrow-shaped ground slot harmonic rejection antenna for 2.45-GHz wireless power transfer applications by suppressing higher-order harmonics using open stub and H-slot in the feed line. Desai et al. [40] proposed a multi-linear polarization reconfigurable antenna consisting of a pair of printed dipole antennas with a plus-shaped geometry. By using shorting pins and a coaxial feed network, 0° and 90° polarization reconfigurabilities were obtained. A maximum radiation direction alignment method was proposed by Zhang et al. [50] for WPT using the unexploited third harmonic generated from rectifiers and reflected back by the rectifier filter. The whole system reuses a hybrid coupler to receive the input power and transmit the third harmonic back simultaneously, leading to a highly compact design for WPT applications.

Zhang et al. [51] proposed a method to exploit the third harmonic generated by rectifiers. A ring coupler was employed to distribute the received fundamental frequency power for rectification by the differential charge pump and couple the third harmonic generation back to its input port. Such third harmonic power was fed back by a dual-band antenna operating at the third harmonic frequency. Hafeez et al. [52] proposed the design procedure of a wireless power transfer (WPT) system based on high-efficiency offset reflector antennas fed by conical horns, and the operating frequency of the system was 6 GHz. The performance of the system was optimized by calibrating the feeding horns and the offset reflector's dimensions to minimize the path and reflection losses. The proposed system utilized a 1-W power transfer over a distance of up to 12 m. A low-profile high-gain slot antenna (Figure 5.9) using a novel metamaterial-inspired superstrate was presented in Ref. [53]. The superstrate was placed at the height of 0.065 λ_0 from the reference antenna. The peak realized gain of the proposed antenna was 8.24 dBi, which was 5.24 dB higher than that of the reference antenna. The reference antenna was an SIW-based slot antenna acting as a magnetic dipole antenna, and it had a peak gain of 3 dBi.

5.3.2 Multi-Band Antennas

Popovic [54] presented an overview of a system for wireless far-field powering of unattended distributed wireless sensors, including the power reception device design and optimization, power transmission, power management, and control. The design methodology for antennas integrated with rectifiers (rectennas) optimized for efficiency at low incident power levels (5–100 μW/cm^2) and implemented in various parts of the system in the 2 GHz cellular and 2.45 GHz unlicensed bands, are discussed in [55]. Chi [56] introduced a novel printed coil integrated with a multi-band antenna. The coil can be used for Qi wireless charger, and the multi-band antenna covers the frequency bands of GSM, GPS, DCS, PCS, UMTS, Wi-Fi, Bluetooth, and WiMAX.

Das et al. [57], [58,59] proposed a triple-band flexible implantable antenna that is tuned by using a ground slot in three specific bands, namely Medical Implant Communication Service (MICS: 402–405 MHz) for telemetry, the midfield band (lower gigahertz: 1.45–1.6 GHz) for WPT, and the industrial, scientific, and medical band (ISM: 2.4–2.45 GHz) for power conservation. A midfield transmitter antenna was designed by using two ports and two short pins at a midfield frequency for WPT in a capsule prototype, and a near-field plate (NFP) was used for the prevention of power leakage and increasing the power transfer efficiency in the medical implant. A T-shaped ground slot was used to tune the antenna, and this antenna was wrapped inside a printed 3D capsule prototype to

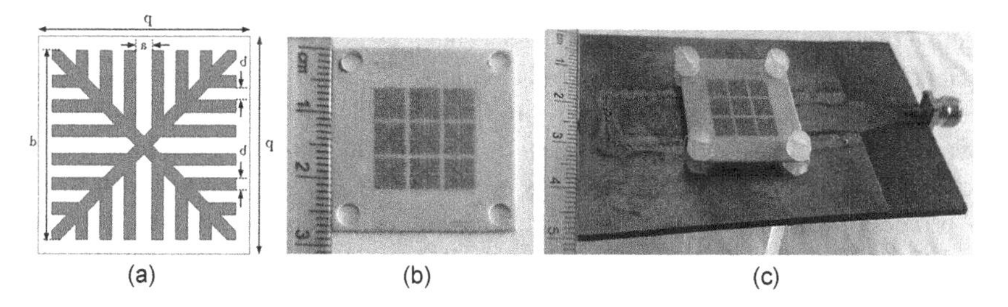

FIGURE 5.9 (a) Metamaterial unit cell; (b) 3×3 unit cells of the MTM; (c) design. (From Pandit, S., Mohan, A., and Ray, P., *Micro. Opt. Technol. Lett.*, 61, 2068–2073, 2019. With permission.)

demonstrate its applicability in different biomedical devices. Belo et al. [60] proposed a solution for a fully passive sensor inside a spaceship, as a WSN powered by WPT technologies. By using this, data and power can be transferred simultaneously, allowing the sensor to be continuously wirelessly powered.

A new miniaturized implantable antenna (MIA) was proposed by Zhang et al. [61] for biomedical applications, which was composed of a loop integrated with split resonate rings (SRRs). It produced two resonant modes to make the antenna operate at both Medical Implant Communication Service (MICS 402–405 MHz) and industrial, scientific, and medical (ISM: 2.40–2.48 GHz) bands. In order to meet the requirements for potential high-data-rate communication, the 2.45-GHz ISM band was utilized to transmit the real-time physiological data. The MICS band was used to transmit the power to improve the lifetime of the implantable system. A new multiband fractal geometry microstrip antenna was developed by Benyetho et al. [62], which was printed over an FR4 substrate with a dimension of 60×30 mm^2. It was suitable for a multi-band rectenna circuit for ISM (industrial, scientific, and medical) band at 2.45 and 5.8 GHz.

Haerinia et al. [63] proposed a dual-band printed planar antenna operating at two ultra-high-frequency bands (2.5/4.5 GHz), which can be used for wireless power transfer in wearable applications. The receiving antenna was printed on a Kapton polyimide-based flexible substrate, and the transmitting antenna was printed on an FR4 substrate. Haerinia et al. [55] proposed a hybrid system including a high-frequency (HF) coil and a dual-UHF antenna for WPT applications. The coils' operating frequency was 510 MHz, and the antennas worked at 2.48 and 4.66 GHz to minimize the system size for small sensors.

5.3.3 Circularly Polarized Antennas

Strassner et al. [64,65] designed a 5.8-GHz circularly polarized (CP) high-efficiency high-gain rectenna. A (CP) dual-rhombic-loop antenna (DRLA) with a gain as high as 10.7 dBi and a VSWR of 2 over a 10% bandwidth was used. A coplanar stripline (CPS) band-reject filter (BRF) was employed to suppress the second harmonic generated by the nonlinear diode and prevent its reradiation. Also, a DC-pass filter was used to minimize the leakage of RF power into the resistive DC load. Circular polarization was chosen for the rectenna to maintain constant output DC power irrespective of its orientation and achieve 82% efficiency at 5.8 GHz.

Heikkinen et al. [66] proposed a novel circularly polarized shorted annular ring-slot rectenna (rectifying antenna) on a 0.5-mm-thick flexible microwave laminate for the powering of batteryless portable devices. In Ref. [67], a 2.45-GHz rectifying antenna (rectenna) using a compact dual circularly polarized (DCP) patch antenna with an RF–DC power conversion part was presented. The dual polarizations were obtained using two crossed slots etched on the ground plane and coupled to a microstrip feed line. Two accesses allow receiving either LHCP or RHCP senses. Due to the coupling feeding technique, no input LPF was needed between the antenna and the rectifier. An electrically small circularly polarized antenna miniaturized from a $\lambda/2$-turnstile antenna by utilizing the top loading and multiple folding techniques was designed by Yoon et al. [68] and applied to near-field WPT as a means of alleviating orientation dependence. A dual-feed circularly polarized antenna with built-in harmonic rejection capabilities was presented by Barrera et al. [69]. The designed antenna can operate at 5.8 GHz and incorporate circular polarization and harmonic rejection in a single design, thus eliminating the need of a low-pass filter. Re et al. [70] presented a novel retrodirective array (RDA) circuit and antenna system for microwave WPT, which deals with an active transmitter module to retrodirect a received tone, with increased power, from a mobile unit and with circularly polarized (CP) radiation. The proposed structure was composed of two circularly polarized antennas with doubly connected hybrid couplers working as a power combiner because of the orthogonal feeding for the two-port antennas. The connection between the antennas and the couplers permits full-duplex operation. Zainol et al. [71] presented a 2.45-GHz harmonic suppression rectangular patch antenna with circular polarization for WPT applications. The harmonic rejection property of the antenna was achieved by creating an LPF on the transmission feed line on the bottom substrate by embedding the U-slot and the symmetrical arm of inverted U-stub. Carlo et al. [72] presented a microstrip patch antenna working at 868 MHz with a fractional impedance bandwidth of 5% and a gain of 4.14 dB, suitable for the radio frequency wireless power transfer and energy harvesting applications. A wireless power link with circular polarization was studied by Liu et al. [73] for far-field wireless power transmission, for which a miniaturized circularly polarized (CP) implantable antenna was designed at 915 ISM band featuring a good miniaturization with the dimensions of $11 \times 11 \times 1.27$ mm^3 by employing the stub loading and capacitive coupling among the stubs. Also a high-gain external CP antenna was fabricated as a transmit antenna, and a simple half-wave rectifier circuit was utilized for WPT measurement.

In Refs. [74,75], an electrically small Huygens circularly polarized (HCP) rectenna whose antenna was directly matched to its rectifying circuit was developed in the ISM band at 915 MHz. The electrically small HCP antenna consisted of four electrically small near-field resonant parasitic (NFRP) elements: two Egyptian axe dipoles (EADs) and two capacitively loaded loops (CLLs). The rectifier was a full-wave rectifying circuit based on HSMS286C diodes. It was integrated with the HCP antenna on its bottom layer via a coplanar stripline (CPS) without occupying any additional space.

5.3.4 Antenna Arrays for WPT

Ren et al. [76] proposed two new circularly polarized retrodirective rectenna arrays, including a 2×2 array and a 4×4 array. A proximity-coupled microstrip ring antenna was used as the retrodirective rectenna array element, which can automatically block harmonic signals up to the third order reradiated by the rectifying circuit. Applications of retrodirective antenna arrays for short-range wireless power transmission were proposed by Li and Jandhyala [77]. They examined the time reversal or phase conjugate theory and design considerations of the array factors in GHz regime and meter range. The wireless power transmission efficiency was analyzed in the 2D case with full-wave EM solution and validated by analytical array theory. A directional antenna using beamforming was proposed by Kim et al. [78] by designing a two-by-two array patch antenna, which was designed to operate at 2.4 GHz and had about 11 times higher gain than a single-patch antenna in any condition. Pookkapund et al. [79] proposed an antenna to operate at 2.45–GHz. The proposed structure consisted of four printed dipoles arranged perpendicularly to one another to form unidirectional pattern for increasing the gain. Ashoor et al. [80] presented 5×5 dielectric resonator antennas (DRAs) for efficient energy harvesting or wireless power transfer in the 5.5-GHz microwaves regime and also highlighted the importance of mutual coupling between adjacent elements on the absorbed power efficiency. Ding et al. [81] focused on the application of advanced smart antenna technologies to simultaneous wireless information and power transfer (SWIPT), including MIMO and relaying techniques to significantly improve the energy efficiency and the spectral efficiency of SWIPT and to achieve a favorable trade-off between system performance and complexity. A low-power wireless power transfer experiment was performed by Park and Kim [82] to efficiently transfer energy using a 4×4 phased array antenna (PAA). A novel compact 4×4 Butler matrix to operate at 2.4 GHz with an array of patch was proposed and designed by Kuek et al. [83] for power transfer to aid electromagnetic energy harvesting for sensors having beam scanning capability of over 100°. Hu et al. [84] demonstrated novel one- and two-dimensional antenna arrays for microwave WPT (MWPT) systems, which can be used as the MWPT receiving antenna of an integrated MWPT and Bluetooth (BLE) communication module (MWPT–BLE module) for smart CNC (computer numerical control) spindle incorporated with the cloud computing system SkyMars. A wireless power transmission array antenna with harmonic suppression, which eliminates the need for a harmonic filter circuit in the rectenna circuit, was designed by Wu et al. [85], in which the second and third harmonics were suppressed to less than –10 dB using the stub on the antenna patch and the slot on the ground plane. Nayeri [86] demonstrated how the focused arrays can significantly increase the total received power at the receiver, and this capability can significantly enhance the efficiency of these wireless energy systems. In Ref. [87], a novel fast and accurate approach for the AoA estimation of lens antenna arrays (LAA) was presented. The key idea was using the difference of the outputs between the adjacent lenses' DFT beams, referred to as "DFT beam difference (DBD)." Therefore, they are more robust against noises and more likely to be correctly identified. It was also proved that the AoA can be accurately estimated based on the signal strengths of the two DBDs, independent of the signal itself. Almorabeti et al. [88] proposed a switched beam smart antenna based on a planar 4×4 Butler matrix in order to improve the performance of wireless power transfer systems to a MAV at 5.8 GHz. The design of the 4×4 planar BM consisted of two crossovers, four 90° hybrid couplers, and two 45° phase shifters. The BM was then used to feed an array formed by four inset-fed microstrip patch antennas in order to produce four beams in desired directions using PIN diode switches. In Ref. [89], a parabolic-shaped retrodirective array was proposed for achieving higher efficiency of microwave power transmission. The parabolic structure was arranged to have the same phase from the receiver, and the efficiency of microwave power transmission in the receiver was higher than that of the equi-phased arrays and retrodirective arrays of linear structure.

REFERENCES

1. M. Shidujaman, H. Samani and M. Arif, "Wireless power transmission trends," *2014 International Conference on Informatics, Electronics & Vision* (ICIEV), Dhaka, pp. 1–6, 2014.
2. X. Lu, P. Wang, D. Niyato, D. I. Kim and Z. Han, "Wireless Networks With RF Energy Harvesting: A Contemporary Survey," in IEEE Communications Surveys & Tutorials, vol. 17, no. 2, pp. 757–789, Secondquarter 2015, doi: 10.1109/COMST.2014. 2368999.
3. Sun-Hee Kim, Yong-Seok Lim, and Seung-Jun Lee, "Magnetic Resonant Coupling Based Wireless Power Transfer System with In-Band Communication," Journal of Semiconductor Technology and Science, Vol.13, No.6, December, 2013 pp. 562–568. HYPERLINK "http://dx.doi.org/10.5573/JSTS.2013.13.6.562" doi:10.5573/JSTS.2013.13.6.562.
4. X. Lu, P. Wang, D. Niyato, D.I. Kim, and Z. Han, "Wireless charging technologies: fundamentals, standards, and network applications," *IEEE Communications Surveys & Tutorials*, vol. 18, no. 2, pp. 1413–1452, Second quarter 2016.
5. N. Keskin and H. Liu, "Unit antenna based wireless power transfer systems," *2015 IEEE 65th Electronic Components and Technology Conference (ECTC)*, San Diego, CA, 2015, pp. 1828–1833. doi: 10.1109/ECTC.2015.7159848.
6. A.P. Sample, D.T. Meyer, and J.R. Smith, "Analysis, experimental results, and range adaptation of magnetically coupled resonators for wireless power transfer," *IEEE Transactions on Industrial Electronics*, vol. 58, no. 2, pp. 544–554, February 2011.
7. H. Lyu, J. Wang, J. La, J.M. Chung, and A. Babakhani, "An energy-efficient wirelessly powered millimeter-scale neuro stimulator implant based on systematic co-design of an inductive loop antenna and a custom rectifier," *IEEE Transactions on Biomedical Circuits and Systems*, vol. 12, no. 5, pp. 1131–1143, October 2018.
8. J. Park, S. Lee, Y. Tak, and S. Nam, "Simple efficient resonant coupling wireless power transfer system operating at varying distances between antennas," *Microwave and Optical Technology Letters*, vol. 54, no. 10, pp. 2397–2401, October 2012.
9. L. Aluigi, T. T. Thai, M. M. Tentzeris, L. Roselli and F. Alimenti, "Chip-to-package wireless power transfer and its application to mm-Wave antennas and monolithic radiometric receivers," *2013 IEEE Radio and Wireless Symposium*, Austin, TX, 2013, pp. 202–204. doi: 10.1109/RWS.2013.6486688
10. H. Hirayama, T. Amano, N. Kikuma and K. Sakakibara, "Undesired emission and biological effect of open-end and short-end antennas for coupled-resonant wireless power transfer," *2013 Asia-Pacific Symposium on Electromagnetic Compatibility (APEMC)*, Melbourne, VIC, 2013, pp. 1–4. doi: 10.1109/APEMC.2013.7360618
11. H. Hirayama, S. Fukasawa, H. Yamada, N. Kikuma and K. Sakakibara, "Open-end and short-end helical antennas for coupled-resonant wireless power transfer," 2014 International Symposium on Electromagnetic Compatibility, Gothenburg, 2014, pp. 1–4, doi: 10.1109/EMCEurope.2014.6930865.
12. K. Fotopoulou and B. W. Flynn, "Wireless powering of implanted sensors using RF inductive coupling," *SENSORS, 2006 IEEE*, Daegu, pp. 765–768, 2006.
13. T. Imura and Y. Hori, "Optimization using transmitting circuit of multiple receiving antennas for wireless power transfer via magnetic resonance coupling," *2011 IEEE 33rd International Telecommunications Energy Conference (INTELEC)*, Amsterdam, 2011, pp. 1–4. doi: 10.1109/INTLEC.2011.6099796.
14. H. Hirayama, T. Amano, N. Kikuma and K. Sakakibara, "A consideration of open- and short-end type helical antennas for magnetic-coupled resonant wireless power transfer," *2012 6th European Conference on Antennas and Propagation (EUCAP)*, Prague, 2012, pp. 3009–3013. doi: 10.1109/EuCAP.2012.6206317

15. T. Imura, "Equivalent circuit for repeater antenna for wireless power transfer via magnetic resonant coupling considering signed coupling," *2011 6th IEEE Conference on Industrial Electronics and Applications*, Beijing, 2011, pp. 1501–1506. doi: 10.1109/ICIEA.2011.5975828
16. M. Taguchi and T. Hirata, "Mutual coupling characteristics of two unbalanced fed ultra low profile inverted L antennas closely faced each other," *2012 IEEE MTT-S International Microwave Workshop Series on Innovative Wireless Power Transmission: Technologies, Systems, and Applications*, Kyoto, 2012, pp. 135–138. doi: 10.1109/IMWS.2012.6215791
17. G. Lee, H. Gwak, Y. Kim and W. Park, "Wireless power transfer system for diagnostic sensor on rotating spindle," *2013 IEEE Wireless Power Transfer (WPT)*, Perugia, 2013, pp. 100–102. doi: 10.1109/WPT.2013.6556892
18. N. Phokhaphan et al., "Wireless power transfer based on MHz inverter through PCB antenna," *2013 1st International Future Energy Electronics Conference (IFEEC)*, Tainan, 2013, pp. 126–130. doi: 10.1109/IFEEC.2013.6687491
19. W. Lee, W. Son, K. Oh, and J. Yu, "Contactless energy transfer systems using antiparallel resonant loops," *IEEE Transactions on Industrial Electronics*, vol. 60, no. 1, pp. 350–359, January 2013.
20. I. Mayordomo, T. Dräger, J. A. Alayón and J. Bernhard, "Wireless power transfer for sensors and systems embedded in fiber composites," *2013 IEEE Wireless Power Transfer (WPT)*, Perugia, 2013, pp. 107–110. doi: 10.1109/WPT.2013.6556894
21. Y. D. Chung, D. W. Kim and S. W. Yim, "Design and performance of wireless power transfer with high temperature superconducting resonance antenna," *2014 IEEE Wireless Power Transfer Conference*, Jeju, 2014, pp. 182–185. doi: 10.1109/WPT.2014.6839577
22. Y. Kim, H. Lee, J. Bang, and C. Chung, "A new design of wireless power transfer system using hellical resonators applicable to multi-channel power transmission," *2014 IEEE Wireless Power Transfer Conference*, Jeju, pp. 277–279, 2014.
23. H. Choi, S. Lee and C. Cha, "Optimization of geometric parameters for circular loop antenna in magnetic coupled wireless power transfer," *2014 IEEE Wireless Power Transfer Conference*, Jeju, 2014, pp. 280–283. doi: 10.1109/WPT.2014.6839563
24. S. Pu, H. T. Hui, C. Liu and Z. Wu, "A new configuration of coil antennas for efficient wireless power transmission systems compatible with different loads," *2014 IEEE Wireless Power Transfer Conference*, Jeju, 2014, pp. 110–113. doi: 10.1109/WPT.2014.6839606
25. Y.G. Kim and S. Nam, "Spherical mode-based analysis of wireless power transfer between two antennas," *IEEE Transactions on Antennas and Propagation*, vol. 62, no. 6, pp. 3054–3063, June 2014.
26. R. Das and H. Yoo, "Wireless power transfer to a pacemaker by using metamaterials and Yagi-Uda antenna concept," *2015 International Workshop on Antenna Technology (iWAT)*, Seoul, 2015, pp. 353–354. doi: 10.1109/IWAT.2015.7365283
27. N. Kajiura and H. Hirayama, "Improvement of transmission efficiency using shielded-loop antenna for wireless power transfer," *2016 International Symposium on Antennas and Propagation (ISAP)*, Okinawa, 2016, pp. 52–53.
28. M. Ando and H. Hirayama, "Impedance matching using folded spiral antenna for coupled-resonant wireless power transfer," *2016 International Symposium on Antennas and Propagation (ISAP)*, Okinawa, 2016, pp. 56–57.
29. K. Nakamura and H. Hirayama, "On a transmission efficiency of tape-wound spiral antenna for coupled resonant wireless power transfer," 2016 International Symposium on Antennas and Propagation (ISAP), Okinawa, 2016, pp. 528–529.

30. H.F. Huang and T. Li, "A spiral electrically small magnetic antenna with high radiation efficiency for wireless power transfer," *IEEE Antennas and Wireless Propagation Letters*, vol. 15, pp. 1495–1498, 2016.
31. T. Ohira and N. Sakai, "Dipole antenna pair revisited from kQ product and Poincare distance for wireless power transfer invited," *2017 IEEE Conference on Antenna Measurements & Applications (CAMA)*, Tsukuba, 2017, pp. 363–366. doi: 10.1109/CAMA.2017.8273453
32. E. Heo, K.Y. Choi, J. Kim, J.H. Park, and H. Lee, "A wearable textile antenna for wireless power transfer by magnetic resonance," *Textile Research Journal*, vol. 88, no. 8, pp. 913–921, February 1, 2017.
33. W. Xiong, M. Jiang, G. Huang, and H. Chen, "Analysis on transfer efficiency of five different antenna configurations in short-distance wireless power transfer," *IEEE International Conference on Electron Devices and Solid State Circuits* (EDSSC), Shenzhen, pp. 1–2, 2018.
34. W. Lu, T. Chu, and T. Wang, "Resonant modeling of wireless power transfer with miniaturized antenna using I type ferrite core in radio propagation for ISM band," *2018 International Workshop on Antenna Technology* (iWAT), Nanjing, pp. 1–3, 2018.
35. H. Zhang, S. Gao, T. Ngo, W. Wu, and Y. Guo, "Wireless power transfer antenna alignment using intermodulation for two-tone powered implantable medical devices," *IEEE Transactions on Microwave Theory and Techniques*, vol. 67, no. 5, pp. 1708–1716, May 2019.
36. T. Shaw and D. Mitra, "Wireless power transfer system based on magnetic dipole coupling with high permittivity metamaterials," *IEEE Antennas and Wireless Propagation Letters*, vol. 18, pp. 1823–1827, 2019.
37. D. Lin, H. Chou, and J. Chou, "Antenna pairing for highly efficient wireless power transmission in the reactive near-field region based on mutual coupled impedance compensation," *IET Microwaves, Antennas & Propagation*, vol. 14, no. 1, pp. 60–65, 2020.
38. T. Kumagai, K. Saito, M. Takahashi, and K. Ito, "A small 915 MHz receiving antenna for wireless power transmission aimed at medical applications," *International Journal of Technology*, vol. 2, pp. 20–27, 2011.
39. Z. Popovic, E. Falkenstein and R. Zane, "Low-power density wireless powering for battery-less sensors," *2013 IEEE Radio and Wireless Symposium*, Austin, TX, 2013, pp. 31–33. doi: 10.1109/RWS.2013.6486631
40. A. Desai and P. Nayeri, "A multi-linear polarization reconfigurable plus shaped dipole antenna for wireless energy harvesting applications," *2018 International Applied Computational Electromagnetics Society Symposium (ACES)*, Denver, CO, 2018, pp. 1–2. doi: 10.23919/ROPACES.2018.8364267
41. D. Takei, K. Saito and K. Ito, "Small antenna stowed in capsular endoscope for wireless power transmission," *2015 International Workshop on Antenna Technology (iWAT)*, Seoul, 2015, pp. 355–356. doi: 10.1109/IWAT.2015.7365284
42. N. Zainol, Z. Zakaria, M. Abu and M. M. Yunus, "Stacked patch antenna harmonic suppression at 2.45 GHz for wireless power transfer," *2015 IEEE International Conference on Control System, Computing and Engineering (ICCSCE)*, George Town, 2015, pp. 96–100. doi: 10.1109/ICCSCE.2015.7482165
43. G. Yang, M.R. Islam, R.A. Dougal, and M. Ali, "A reconfigurable stacked patch antenna for wireless power transfer and data telemetry in sensors," *Progress in Electromagnetics Research C*, vol. 29, 67–81, 2012.
44. B. Wang, H. Liu, C. Liu, L. Zhang, J. Pan and Y. Okuno, "A UHF RFID reader antenna with near field inductive wireless power transfer feature," *2016 Progress in Electromagnetic Research Symposium (PIERS)*, Shanghai, 2016, pp. 1319–1322. doi: 10.1109/PIERS.2016.7734646

45. K. Matsui et al., "Microstrip antenna and rectifier for wireless power transfar at 94GHz," *2017 IEEE Wireless Power Transfer Conference (WPTC)*, Taipei, 2017, pp. 1–3. doi: 10.1109/WPT.2017.7953902
46. G. Dias, P. Pinho, R. Gonçalves and N. Carvalho, "3D antenna for wireless power transmission: Aperture coupled microstrip antenna with dielectric lens," *2017 International Applied Computational Electromagnetics Society Symposium - Italy (ACES)*, Florence, 2017, pp. 1–2. doi: 10.23919/ROPACES.2017.7916328
47. R. Gonçalves, P. Pinho and N. B. Carvalho, "3D printed lens antenna for wireless power transfer at Ku-band," *2017 11th European Conference on Antennas and Propagation (EUCAP)*, Paris, 2017, pp. 773–775. doi: 10.23919/EuCAP.2017.7928092
48. S. B. Belamgi, S. Ray and P. Das, "Suspended planar patch antenna for wireless energy transfer at 2.45GHz," *International Conference on Electronics, Communication and Instrumentation (ICECI)*, Kolkata, 2014, pp. 1–4. doi: 10.1109/ICECI.2014.6767381
49. A. Kumar, U. Pattapu, A. C. Das and S. Das, "A 2.45 GHz harmonic rejection antenna for wireless power transfer applications," *2017 IEEE-APS Topical Conference on Antennas and Propagation in Wireless Communications (APWC)*, Verona, 2017, pp. 81–84, doi: 10.1109/APWC.2017.8062247.
50. H. Zhang, Y.X. Guo, S.P. Gao, and W. Wu, "Wireless power transfer antenna alignment using third harmonic," *IEEE Microwave and Wireless Components Letters*, vol. 28, no. 6, pp. 536–538, June 2018.
51. H. Zhang, Y.X. Guo, S.P. Gao, and W. Wu, "Exploiting third harmonic of differential charge pump for wireless power transfer antenna alignment," *IEEE Microwave and Wireless Components Letters*, vol. 29, no. 1, pp. 71–73, January 2019.
52. M.A. Hafeez, K. Yousef, M. AbdelRaheem, and E.E.M. Khaled, "Design of 6 GHz high efficiency long range wireless power transfer system using offset reflectors fed by conical horns," *2019 International Conference on Innovative Trends in Computer Engineering* (ITCE), Aswan, Egypt, pp. 365–370, 2019.
53. S. Pandit, A. Mohan, and P. Ray, "Metamaterial-inspired low-profile high-gain slot antenna," *Microwave and Optical Technology Letters*, vol. 61, pp. 2068–2073, 2019.
54. Z. Popovic, "Far-field wireless power delivery and power management for low-power sensors," *2013 IEEE Wireless Power Transfer (WPT)*, Perugia, 2013, pp. 1–4. doi: 10.1109/WPT.2013.6556867
55. M. Haerinia and S. Noghanian, "Design of Hybrid Wireless Power Transfer and Dual Ultrahigh-Frequency Antenna System," *2019 URSI International Symposium on Electromagnetic Theory (EMTS)*, San Diego, CA, USA, 2019, pp. 1–4. doi: 10.23919/URSI-EMTS.2019.8931514
56. Y. Chi, "Integration of coil and UHF antenna for wirless wireless power transfer and data telemetry," *2016 IEEE Wireless Power Transfer Conference (WPTC)*, Aveiro, 2016, pp. 1–3. doi: 10.1109/WPT.2016.7498831
57. R. Das, Y. Cho and H. Yoo, "High efficiency unidirectional wireless power transfer by a triple band deep-tissue implantable antenna," *2016 IEEE MTT-S International Microwave Symposium (IMS)*, San Francisco, CA, 2016, pp. 1–4. doi: 10.1109/MWSYM.2016.7540252
60. D. Belo, R. Correia, F. Pereira and N. B. De Carvalho, "Dual band wireless power and data transfer for space-based sensors," *2017 Topical Workshop on Internet of Space (TWIOS)*, Phoenix, AZ, 2017, pp. 1–4. doi: 10.1109/TWIOS.2017.7869773
58. R. Das and H. Yoo, "A triple-band deep-tissue implantable antenna incorporating biotelemetry and unidirectional wireless power transfer system," *2017 IEEE International Symposium on Antennas and Propagation & USNC/URSI National Radio Science Meeting*, San Diego, CA, 2017, pp. 2489–2490. doi: 10.1109/APUSNCURSINRSM.2017.8073287

59. R. Das and H. Yoo, "A Multiband Antenna Associating Wireless Monitoring and Nonleaky Wireless Power Transfer System for Biomedical Implants," in *IEEE Transactions on Microwave Theory and Techniques*, vol. 65, no. 7, pp. 2485–2495, July 2017. doi: 10.1109/TMTT.2017.2647945
61. H. Zhang, L. Li, C. Liu, Y.X. Guo, and S. Wu, "Miniaturized implantable antenna integrated with split resonate rings for wireless power transfer and data telemetry," *Microwave and Optical Technology Letters*, vol. 59, no. 3, pp. 710–714, 2017.
62. T. Benyetho, J. Zbitou, L. El Abdellaoui, H. Bennis, and A. Tribak, "A new fractal multiband antenna for wireless power transmission applications," *Active and Passive Electronic Components*, vol. 2018, Article ID 2084747, pp. 10, 2018.
63. M. Haerinia and S. Noghanian, "A printed wearable dual-band antenna for wireless power transfer," *Sensors*, vol. 19, pp. 1732, 2019. doi: 10.3390/s19071732.
64. B. Strassner and K. Chang, "5.8-GHz circularly polarized rectifying antenna for wireless microwave power transmission," *IEEE Transactions on Microwave Theory and Techniques*, vol. 50, no. 8, pp. 1870–1876, August 2002.
65. B. Strassner and K. Chang, "5.8-GHz circularly polarized dual-rhombic-loop traveling-wave rectifying antenna for low power-density wireless power transmission applications," *IEEE Transactions on Microwave Theory and Techniques*, vol. 51, no. 5, pp. 1548–1553, May 2003.
66. J. Heikkinen and M. Kivikoski, "Low-profile circularly polarized rectifying antenna for wireless power transmission at 5.8 GHz," *IEEE Microwave and Wireless Components Letters*, vol. 14, no. 4, pp. 162–164, April 2004.
67. Z. Harouni, L. Cirio, L. Osman, A. Gharsallah, and O. Picon, "A dual circularly polarized 2.45-GHz rectenna for wireless power transmission," *IEEE Antennas and Wireless Propagation Letters*, vol. 10, pp. 306–309, 2011.
68. I.J. Yoon and H. Ling, "Design of an electrically small circularly polarised turnstile antenna and its application to near-field wireless power transfer," *IET Microwaves, Antennas & Propagation*, vol. 8, pp. 308–314, 2013.
69. O.A. Barrera, D.H. Lee, N.M. Quyet, and H.C. Park, "A circularly polarized harmonic-rejecting antenna for wireless power transfer applications," *IEICE Electronics Express*. 10. 20130665–20130665. doi: 10.1587/elex.10.20130665.
70. P. D. H. Re, S. K. Podilchak, C. Constantinides, G. Goussetis and J. Lee, "An active retrodirective antenna element for circularly polarized wireless power transmission," *2016 IEEE Wireless Power Transfer Conference (WPTC)*, Aveiro, 2016, pp. 1–4. doi: 10.1109/WPT.2016.7498805
71. N. Zainol, Z. Zakaria, M. Abu, and M.M. Yunus, "A 2.45 GHz harmonic suppression rectangular patch antenna with circular polarization for wireless power transfer application," *IETE Journal of Research*, vol. 10, pp. 310–316, 2017.
72. C.A. Di Carlo, L. Di Donato, G.S. Mauro, R. La Rosa, P. Livreri, and G. Sorbello, "A circularly polarized wideband high gain patch antenna for wireless power transfer," *Microwave and Optical Technology Letters*, vol. 60, pp. 620–625, 2018.
73. C. Liu, Y. Zhang, and X. Liu, "Circularly polarized implantable antenna for 915 MHz ISM-band far-field wireless power transmission," *IEEE Antennas and Wireless Propagation Letters*, vol. 17, no. 3, pp. 373–376, March 2018.
74. W. Lin and R. W. Ziolkowski, "Electrically Small, Highly Efficient, Huygens Circularly Polarized Rectenna for Wireless Power Transfer Applications," *2019 13th European Conference on Antennas and Propagation (EuCAP)*, Krakow, Poland, 2019, pp. 1–3.
75. W. Lin and R.W. Ziolkowski, "Electrically small Huygens CP rectenna with a driven loop element maximizes its wireless power transfer efficiency," *IEEE Transactions on Antennas and Propagation*, vol. 68, no. 1, pp. 540–545, January 2019.

76. Y.J. Ren and K. Chang, "New 5.8-GHz circularly polarized retrodirective rectenna arrays for wireless power transmission," *IEEE Transactions on Microwave Theory and Techniques*, vol. 54, pp. 2970–2976, 2006.
77. Y. Li, and V. Jandhyala, "Design of retrodirective antenna arrays for short-range wireless power transmission," *IEEE Transactions on Antennas and Propagation*, vol. 60, no. 1, pp. 206–211, January 2012.
78. J. Kim, H. Yu, and C. Cha, "Efficiency enhancement using beam forming array antenna for microwave-based wireless energy transfer," *2014 IEEE Wireless Power Transfer Conference*, Jeju, pp. 288–291, 2014.
79. K. Pookkapund, K. Boonying, S. Dentri, and C. Phongcharoenpanich, "Planar array antenna for WPT system at 2.4 GHz," *2014 International Symposium on Antennas and Propagation Conference Proceedings*, Kaohsiung, pp. 541–542, 2014.
80. A.Z. Ashoor and O.M. Ramahi, "Dielectric resonator antenna arrays for microwave energy harvesting and far-field wireless power transfer," *Progress in Electromagnetics Research C*, vol. 59, pp. 89–99, 2015.
81. Z. Ding, C. Zhong, D.W.K. Ng, M. Peng, H.A. Suraweera, R. Schober, and H.V. Poor, "Application of smart antenna technologies in simultaneous wireless information and power transfer," *IEEE Communications Magazine*, vol. 53, pp. 86–93, April 2015.
82. H. Park and H. Kim, "Effective wireless low-power transmission using a phased array antenna," *Microwave and Optical Technology Letters*, vol. 57, no. 2, pp. 421–424, February 2015.
83. J. J. Kuek, K. T. Chandrasekaran, M. F. Karim, Nasimuddin and A. Alphones, "A compact Butler matrix for wireless power transfer to aid electromagnetic energy harvesting for sensors," *2017 IEEE Asia Pacific Microwave Conference (APMC)*, Kuala Lumpar, 2017, pp. 334–336. doi: 10.1109/APMC.2017.8251447
84. C. Hu, Y. Lin, C. Chang, P. Tsao, K. Lan and C. Yeh, "One- and two-dimensional antenna arrays for microwave wireless power transfer (MWPT) systems," *2017 IEEE Wireless Power Transfer Conference (WPTC)*, Taipei, 2017, pp. 1–5. doi: 10.1109/WPT.2017.7953903
85. C. Wu, G. Pan, H. Hsu and J. Sun, "A 2.45-GHz planar array antenna with harmonic suppression for wireless power transmission applications," *2017 IEEE Wireless Power Transfer Conference (WPTC)*, Taipei, 2017, pp. 1–3. doi: 10.1109/WPT.2017.7953835
86. P. Nayeri, "Focused antenna arrays for wireless power transfer applications," *2018 International Applied Computational Electromagnetics Society Symposium (ACES)*, Denver, CO, 2018, pp. 1–2. doi: 10.23919/ROPACES.2018.8364266
87. K. Wu, W. Ni, T. Su, R. P. Liu, and Y. J. Guo, "Efficient angle-of-arrival estimation of lens antenna arrays for wireless information and power transfer," *IEEE Journal on Selected Areas in Communications*, vol. 37, no. 1, pp. 116–130, January 2019.
88. S. Almorabeti, M. Rifi, H. Terchoune, and H. Tizyi, "Design and implementation of a switched beam smart antenna for wireless power transfer system at 5.8 GHz," *2018 Renewable Energies, Power Systems & Green Inclusive Economy* (REPS-GIE), Casablanca, pp. 1–6, 2018.
89. S. Kim, J.W. Kim, J.W. Kim, G. Kim, and J.W. Yu, "Retro-directive array antenna with parabolic shape structure for short-range microwave power transfer," *2019 IEEE International Symposium on Antennas and Propagation and USNC-URSI Radio Science Meeting*, Atlanta, GA, pp. 1799–1800, 2019.

6 Rectifiers for WPT Systems

6.1 INTRODUCTION

Wireless power transfer (WPT) techniques have many applications in consumer electronics, industries, transportation, implantable medical devices, etc. [1]. Nowadays, more and more devices use WPT for powering or charging battery driven wireless devices due to the flexible charging tract [2]. Wireless powering through inductive link is used in biomedical implants to avoid the battery replacement issues [3]. But, the output of the source and the input of the receiver in wireless power transfer systems are an AC voltage that requires conversion into DC by a rectifier at the receiver end. The efficiency of the rectifier to covert AC to DC highly impacts on the performance of the WPT receiver. Hence, there is a trend among researchers to design highly efficient rectifiers for WPT systems for different applications. Several studies are available in the literature discussing the features of different types of rectifiers, such as active rectifier, reconfigurable rectifier, resonant regulating rectifier, CMOS-based rectifier, and class-based rectifier, respectively.

6.2 CLASS-BASED RECTIFIERS

Inductively coupled power transfer (IPT) systems are commonly used for wireless powering. Hence, to obtain high power transfer efficiency, it is necessary to design efficient transmitters and rectifiers [4]. The rectifier plays a vital role in an IPT system at the receiver end because it is the final power stage and hence can have a significant influence on the overall execution of the system [5]. In general, most of the WPT systems use hard-switching-based rectifiers (full-bridge rectifiers). But, these hard-switching-based topologies have a considerable switching loss at MHz range. Therefore, it is desirable to use the soft-switching topology for rectifiers working at MHz range [7]. The promising candidates for rectification at high frequency are Class-D and Class-E rectifiers because they enable zero-voltage switching (ZVS) or zero-current switching (ZCS) and can reduce the switching loss. In the arrangement of zero-current and zero-voltage switching methods, the current and voltage waveforms come to a zero-current and zero-voltage condition so that the device turns on/off at the proper time instant. Many studies in the literature have discussed the Class-E rectifiers [4,7], hybrid Class-E rectifier [8], Class-D rectifiers [5,9], and Class-DE rectifier [10]. Several other class-based rectifier designs [11–37] are discussed here in this section.

In Ref. [11], a new type of high-frequency high-efficiency resonant DC–DC converter (Class-E^2 converter) based on Class-E inverter and Class-E rectifier was proposed, as shown in Figure 6.1. Inoue et al. [4] proposed an inductively coupled WPT system with a Class-DE transmitter and a Class-E rectifier along with numerical

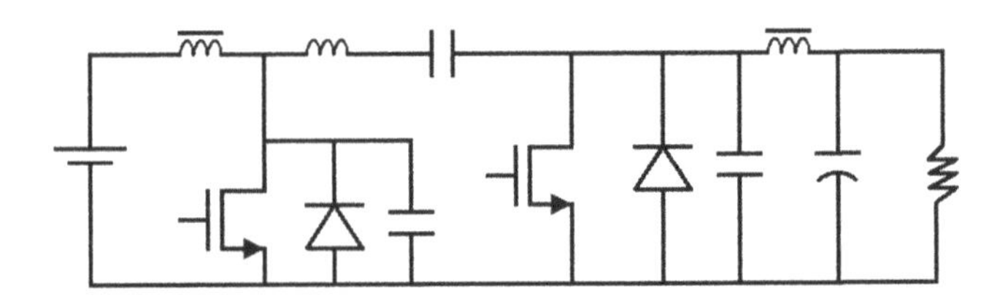

FIGURE 6.1 Synchronous rectifier converter. (From Kazimierczuk, M.K. and Jozwik, J., *IEEE Trans. Indus. Electron.*, 36, 468–478, 1989. With permission.)

design procedure. The proposed system achieved an overall power-transfer efficiency of around 90.4% with 100 W output at 500 kHz operating frequency. Nagashima et al. [7] presented a resonant inductive coupling WPT (RIC-WPT) system with a Class-DE inverter used as a transmitter and a Class-E rectifier used as a receiver along with its analytical design procedure, which achieved a high power conversion efficiency (79.0% overall efficiency at 5 W (50 Ω) output power, coupling coefficient 0.072, and 1 MHz operating frequency).

Luk et al. [5] have presented a class E driven WPT system (Figure 6.2) using half-wave Class D voltage-switching rectifier in which the measured load power was approximately at 10 W and also the efficiency of the Class D rectifier was approximately measured as 91.2%. Liu et al. [6] presented a compact Class-E rectifier for a 6.78-MHz WPT system, achieving a 92% efficiency with an input power of 10 W. Nagashima et al. [10] presented a RIC-WPT system with a Class-E inverter as a transmitter and a Class-DE rectifier as a receiver for achieving a high PCE (73.0% at 9.87 W (50 Ω) output power, coil distance 10 cm, and 1 MHz operating frequency) utilizing the Class-E ZVS/ZDS conditions in both the inverter and the rectifier. In Ref. [9], two current-driven half-wave rectifier topologies (Figure 6.3), Class-D and Class-E, for 6.78-MHz high-power IPT applications were proposed with efficiencies of 95% (Class-D) and 90% (Class-E).

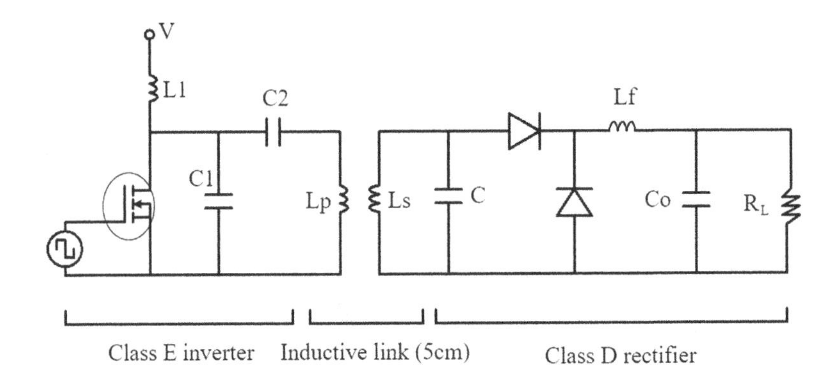

FIGURE 6.2 IPT system with a class-based driver and rectifier. (From Luk, P.C.K. and Aldhaher, S., "Analysis and design of a Class D rectifier for a Class E driven wireless power transfer system," *IEEE Energy Conversion Congress and Exposition* (ECCE), Pittsburgh, PA, 851–857, 2014. With permission.)

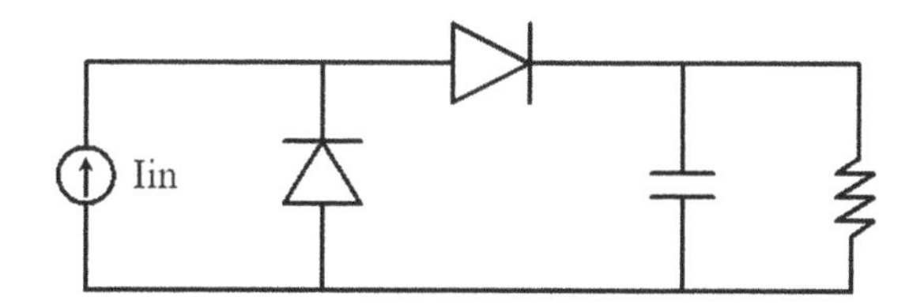

FIGURE 6.3 Class-D half-wave rectifier. (From Kkelis, G., Yates, D.C., and Mitcheson, P.D., "Comparison of current driven Class-D and Class-E half-wave rectifiers for 6.78 MHz high power IPT applications," *IEEE Wireless Power Transfer Conference* (WPTC), Boulder, CO, 1–4, 2015. With permission.)

In Ref. [12], a three-phase resonant inverter for WPT applications was presented. In Ref. [13], a 6.78-MHz WPT system including a Class-E PA, coupling coils, a Class-E current-driven rectifier, and an electronic load was proposed, achieving an efficiency of 84% at a power level of 20 W (Figure 6.4). Aldhaher et al. [14] presented the design and implementation of a Class-EF2 inverter and a Class-EF2 rectifier for two WPT systems, one operating at 6.78 MHz and the other at 27.12 MHz. The peak DC-to-DC efficiency of the 6.78-MHz WPT system was 71% at 29 W load power, and the peak DC-to-DC efficiency of the 27.12-MHz WPT system was 75% at 25 W load power (Figure 6.5).

Rizo et al. [15] designed a GaN HEMT Class-E power amplifier and a Schottky diode Class-E rectifier for the simultaneous transmission of signal and power to a remote receiver for achieving a peak efficiency value of 74% for an input power of 23 dBm. In Ref. [16], a 6.78-MHz WPT system using a power amplifier, a coupling

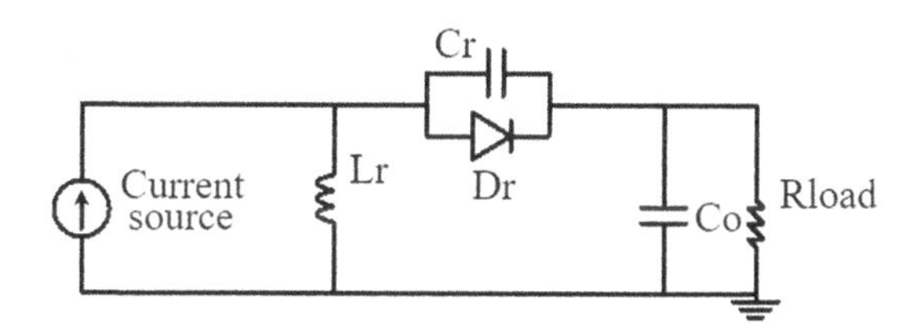

FIGURE 6.4 Class-E current-driven rectifier. (From Liu, M., Fu, M., and Ma, C., *IEEE Trans. Power Electron.*, 31, 4280–4291, 2016. With permission.)

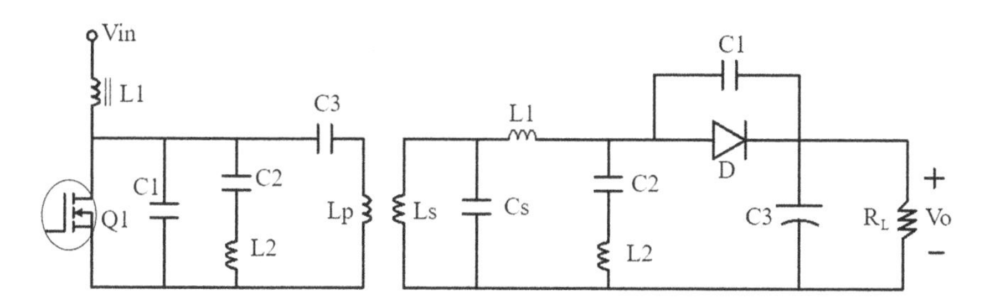

FIGURE 6.5 Circuit of the WPT system. (From Aldhaher, S., Yates, D.C., and Mitcheson, P.D., *IEEE Trans. Power Electron.*, 31, 8138–8150, 2016. With permission.)

coil, and a Class-E rectifier was proposed. In Ref. [8], a hybrid Class-E low-dv/dt rectifier for high-frequency inductive power transfer was proposed.

Liu et al. [17], investigated the application of a Class-E full-wave current-driven rectifier in WPT systems and the performances of both the Class-E rectifier and the conventional full-bridge rectifier as shown in Figure 6.6. The efficiencies of both the rectification (over 91%) and the overall system (around 80%) were obviously improved compared to the system using the conventional full-bridge rectifier. Inaba et al. [18] proposed a WPT system with a Class-E inverter, a half-bridge Class-DE rectifier, and a loosely coupling transformer, in which both the inverter and the rectifier achieve zero-voltage switching (ZVS) and zero-voltage derivative switching (ZVDS) at an optimum coupling coefficient for achieving a power conversion efficiency of 70.99% with the output power of 2.320 W.

Fu et al. [19] presented a design procedure of a Class-E^2 DC–DC converter for MHz WPT systems based on a compact Class-E current-driven rectifier and achieved an optimal efficiency of 85%. Zhang et al. [20] presented a magnetic coupling resonant WPT (MCR-WPT) system and achieved a 79.6% overall efficiency at 3 W output power at 1 MHz operating frequency.

In Ref. [21], a high-efficiency/high-output power, low-noise MHz WPT system over a wide range of mutual inductance based on Class-E PA and Class-E rectifier was proposed for reducing the total harmonic distortion in the input voltage of the coupling coils from 52.9% to 9.6%. In Ref. [22], 6.78-MHz Class-E^2 DC–DC converters with and without a fixed/tunable IMN were presented for providing a stable output power. In Ref. [23], a WPT system with an asymmetrical duty-cycle-controlled Class-D zero-voltage switching (ZVS) inverter and a Class-E rectifier was presented, which allows a high PCE at high frequencies.

In Ref. [24], a Class-E resonant full-wave rectifier in a capacitive power transfer (CPT) system, is proposed with 98% measured efficiency of the designed full-wave Class-E rectifier achieved with a 100 Ω load. In Ref. [25], a soft switching based Class-E power amplifier (PA) and rectifier for 6.78 MHz WPT with low harmonic distortion and high efficiency (70%) is proposed for maintaining high efficiencies.

In Ref. [26], a 6.78-MHz Class-E receiver featuring a soft-switching and temperature-compensated GaN-based active diode (GAD) rectifier was proposed.

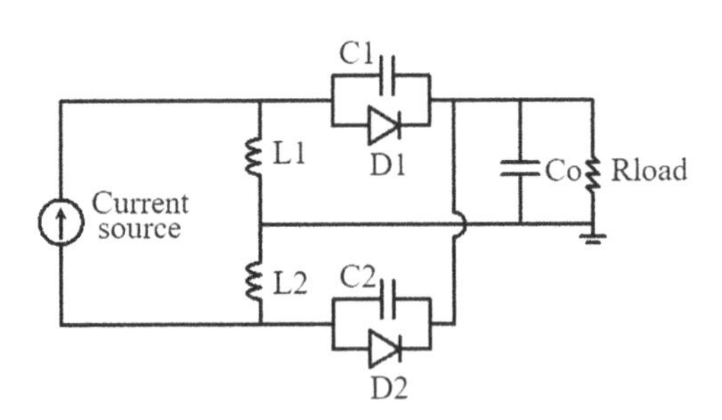

FIGURE 6.6 Class-E rectifier. (From Liu, M., Fu, M., and Ma, C., *IEEE Trans. Power Electron.*, 32, 1198–1209, 2017. With permission.)

Zhu et al. [27] presented a two-stage adaptive impedance matching rectifying circuit delivering an output of 10 W at 100 MHz and achieved a conversion efficiency of 65%.

Zhang et al. [28] analyzed and compared the efficiency of three rectifiers in highly resonant WPT systems: full-bridge (FB) rectifier, voltage doubler (VD), and current-mode Class-D (CD) rectifier and achieved a total system efficiency of 83% at 45 W output power, 80% at 26 W, and 87% at 58 W, respectively. Bati et al. [29] proposed a Class-E^2 converter operating at the frequency of 200 kHz, and it consisted of an inductive link whose primary coil was driven by a Class-E inverter and secondary coil was voltage-driven by a Class-E synchronous rectifier for achieving an excellent efficiency (90%–98%). In Ref. [30], a high-efficiency inverse Class-F microwave rectifier for wireless power transmission was described.

6.3 CMOS RECTIFIERS

The traditional rectifiers are developed using diodes having large voltage drops and hence are not suitable for low-voltage circuits [31]. Diode-connected transistors are also affected by the threshold voltage and the instantaneous voltage drop across it [32]. The other way of treating with this issue is to utilize low-threshold-voltage transistors, but it increases the fabrication cost [31]. A less expensive solution to this problem is provided by CMOS rectifiers, which not only solve the issues of voltage drop, but also give a miniaturized solution. These features also allow the rectifier to be used in the implantable medical devices and other low-voltage applications. CMOS gate cross-coupled rectifiers have the advantages of low on-resistance in comparison with diode-based rectifiers and can achieve average efficiency at low input signals, but have the risk of reverse leakage current [3]. Comparator-based CMOS rectifiers were adopted to minimize the reverse leakage current and maximize the forward current. Guo et al. [33] presented an efficiency-enhanced integrated full-wave CMOS rectifier for the transcutaneous power transmission in high-current biomedical implants for achieving a peak voltage conversion ratio of 95% and the power efficiency of at least 82%.

In Ref. [34], a design of an active back-telemetry rectifier (ABTR) operating at f_{IN} = 13.56 MHz within the ISM band was proposed for achieving a PCE of ~83% and a voltage conversion ratio >92% in post-layout simulations. In Ref. [35], a highly efficient fully integrated passive CMOS rectifier was proposed, and it provided a voltage transmission efficiency of higher than 90% for a wide range of AC input signal amplitude. In Ref. [3], a highly efficient CMOS rectifier for WPT operating at an ISM band of 13.56 MHz was proposed, and it can be used in biomedical implant applications for attaining a high PCE at a small input power. A power-efficient, low-complexity rectifier providing a measured efficiency of 85.8% at 40.68 MHz while delivering 1 mW of power to a biosensor immersed in blood was presented in Ref. [31]. A new full-wave CMOS rectifier dedicated for wirelessly powered low-voltage biomedical implants employing a MOS-based gate cross-coupled nMOS switches and pMOS switches, is proposed in Ref. [36]. Ha et al. [37] presented a rectifier with a comparator using unbalanced body biasing for WPT in a 0.11-μm RF CMOS process. In Ref. [38], a high-efficiency energy harvester comprising a passive voltage-boosting network (VBN) and an orthogonally switching charge pump rectifier (CPR) was presented.

A high-efficiency comparator-based CMOS rectifier for 13.56-MHz WPT biomedical implant devices was designed in Ref. [39]. In Ref. [40], a solution to the inefficient rectification problem was presented by proposing an integrated voltage-controlled-oscillator-based rectifier, efficiently converting 900-MHz RF signals to a DC supply to power up a passive microwatt implantable sensor.

In Ref. [41], a CMOS differential-drive (DDR) full-wave rectifier was optimized for improvement in PCE (83.3%) at –16.3 dBm input power, with an input RF of 953 MHz and 0.6 V peak sinusoidal signal, and the voltage conversion ratio became 72.3% at –13 dBm input power. The CMOS differential-drive full-wave rectifier topology is shown in Figure 6.2. In Ref. [42], a CMOS full-wave rectifier with comparator-controlled switches for wireless transmission of power to implantable medical devices was presented, to achieve 85.5% PCE and 88% voltage conversion efficiency (VCE). A rectifier topology was designed in Ref. [43] to maximize the PCE focusing on minimizing the input voltage to less than 200 mV. Ehsanbakhsh et al. [44] proposed a method for optimizing the Villard voltage multiplier, resulting in enhanced output voltage in comparison with conventional counterparts.

6.4 ACTIVE RECTIFIERS

On the power front, it is imperative for the rectifier to be extremely efficient to convert the small received AC power to DC and present it to the load with minimum dissipation. A more efficient rectifier can deliver a given amount of power for a lesser voltage induced across the secondary coil. Because diode parasitic elements cause increasingly non-ideal behavior as frequency rises, active rectifiers have been proven to be more efficient compared to the diode-connected passive counterparts [45]. Active rectifiers are used to improve the efficiency by using actively controlled switching components such as MOS transistor. To improve the power conversion efficiency of the conventional full-bridge rectifier, a comparator-based active rectifier was proposed [46]. However, the comparator-based active rectifier has leakage problem at high-frequency operation caused due to the turn-on and turn-off delay of main power transistors in the rectifier [47]. Active rectifiers with self-switching comparator technology [48] reduce the static power consumption of the comparators. Some active rectifiers with digital offset compensation [49], off delay compensation [50], on/off delay compensation [51–54], adaptive time delay control [55–57], self-switching comparator [58], and dynamic controllable comparator [58] were also proposed with some special features. In Ref. [47], a novel synchronous rectifier (SR) driving scheme (Figure 6.7) for resonant converters was presented for high-frequency, high-efficiency, and high-power-density DC–DC resonant converters.

In Ref. [1], a highly efficient, high-speed comparator-based active rectifier fabricated using 0.35-μm CMOS technology was proposed, achieving 92% PCE. In Ref. [2], a simple digital control scheme for the active rectifier on the secondary side was proposed to improve the efficiency in comparison with a passive and semi-active rectifier solution. Chinthavali et al. [59] presented a novel approach to the system model and the impact of different control parameters on the load power.

Tang et al. [60] presented a single PFC active rectifier with a voltage boost capability to serve as a grid interface power electronic converter for the WPT applications.

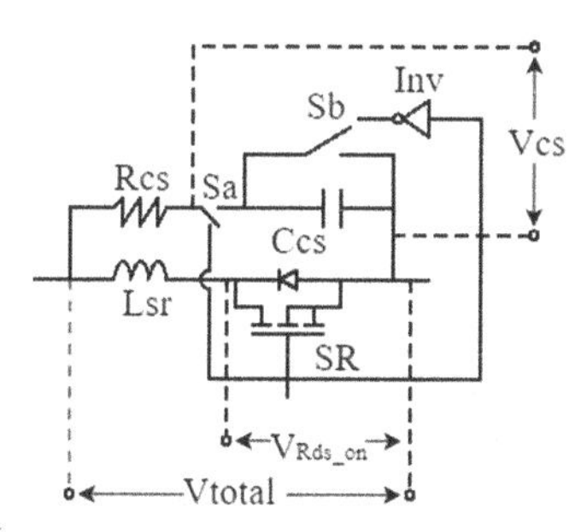

FIGURE 6.7 SR driving scheme. (From Fu, D., Liu, Y., Lee, F.C., and Xu, M., *IEEE Trans. Power Electron.*, 24, 1321–1329, 2009. With permission.)

In Ref. [50], four diodes of a conventional passive rectifier were replaced by two cross-coupled PMOS transistors and two comparator-controlled NMOS switches to eliminate diode voltage drops such that a high voltage conversion ratio and power conversion efficiency could be achieved even at low AC input amplitude. Colak et al. [61] proposed phase control principle of an S-BAR circuit topology for WPT, in which the standard receiver-side rectifier topology was developed by replacing rectifier lower diodes with synchronous switches controlled by a phase-shifted PWM signal.

In Ref. [62], a concept to maximize the efficiency of a WPT system and to increase the amount of extracted power was proposed. In Ref. [63], a 13.56-MHz power-efficient CMOS active rectifier was proposed for implantable medical devices for achieving 92.9% PCE with wide input voltage range. Park et al. [64] presented an inductive coupling wireless power receiver with a high-efficiency active rectifier and a multi-feedback LDO regulator for achieving a maximum PCE of 94.2% of the active rectifier when the load current was 800 mA. In Ref. [65], a full-CMOS receiver was presented to achieve an efficiency of 94.2% for a 6.78-MHz A4WP application. Cochran et al. [45] examined the potential of a GaN-based synchronous rectifier as a receiver in 6.78-MHz WPT applications. In Ref. [66], an on/off delay compensation technique was proposed to eliminate the propagation delays of comparators and gate drivers adaptively and to get a VCR higher than 90% and a PCE higher than 89.1% for a load resistor of 500 Ω. Kim et al. [67] proposed a novel selective harmonic elimination (SHE) method of the radiation noise from an automotive WPT system to reduce the possible EMI caused by the high dv/dt of a resonant inverter and that of a rectifier. The phase-shift control was applied for this in both the inverter and the rectifier. In Ref. [68], a full-wave active rectifier with zero-current sensing technique, fabricated using 0.18-μm CMOS process, was presented for achieving a maximum simulated PCE of 96.8% at 150 kHz. In Ref. [69], a near-optimum active rectifier was proposed to achieve well-optimized PCE and voltage conversion ratio (VCR) under various process, voltage, temperature (PVT), and loading conditions. In Refs. [46,70] authors presented one active rectifier using CMOS transistors, connected as diodes and controlled by two original comparators that are responsible for turning on–off the circuit for achieving 81% efficiency at the output of the rectifier. In Ref. [40], a full active rectifier consisting of GaN devices and a CMOS controller was designed for wireless power transmission in high-power consumer devices, with a peak power transfer efficiency of 91.8%. Xue et al. [56] proposed a novel active rectifier with a delay time controller to solve the issues in active rectifier with comparators (CMPs) and achieved a peak PCE of 95.3% at 200 Ω load.

In Ref. [71], a highly integrated three-level single-inductor dual-output DC–DC cooperating with a 2X active rectifier (voltage doubler) was proposed for the 6.78-MHz WPT RX and was measured with PCB coils. A 13.56-MHz 64.8-mW fully integrated CMOS active rectifier for biomedical wireless power transfer systems was presented [52], and it provided a voltage conversion ratio higher than 90% and a power conversion efficiency higher than 89.1% for a load resistor of 500 Ω. In Ref. [72], the timing of the active rectification mode of bidirectional UWPT systems and its effects on PCE were focused. The work presented in Ref. [73] described the development of a hybrid Class-E synchronous rectifier for wireless powering of quad-copters through inductive power transfer. Hata et al. [74] aimed to achieve power control and efficiency maximization simultaneously without communication between the primary side and the secondary side of the system. In Ref. [75], a 6.78-MHz synchronous rectifier was proposed, as shown in Figure 6.8, to address the issues of total harmonic distortion (THD), parasitic conduction losses, input phase control, and dynamic loading.

Koyama et al. [76] proposed a novel simple self-driven synchronous rectifier (shown in Figure 6.9) for the RIC-WPT to the small mobile apparatus. In Ref. [77], a 13.56-MHz regulated dual-output active rectifier was proposed for implantable medical devices (IMDs) for getting a maximum PCE of 78.5%. In Ref. [78], a 13.56-MHz 50-W load-independent synchronous Class-E rectifier using GaN devices for a wireless charging system was presented. The design of an active rectifier with the inclusion of a shorting control for overvoltage protection was presented in Ref. [79] for wirelessly powered medical implants using 180-nm HV-CMOS technology to achieve an efficiency of greater than 90% with output powers from 40 to 600 mW. Mao et al. [53] proposed a 6.78-MHz active voltage doubler with near-optimal on/off delay compensation for WPT systems. A 13.56-MHz active rectifier for WPT systems with self-switching comparator technology was proposed in Ref. [48] to achieve a peak PCE of 75.4% under 10 dBm input power.

In Ref. [80], a 13.56-MHz active rectifier with a dynamically controllable common-gate comparator for wirelessly powered implantable medical devices was presented for achieving 91.9% PCE and 86%–96.4%voltage conversion efficiency.

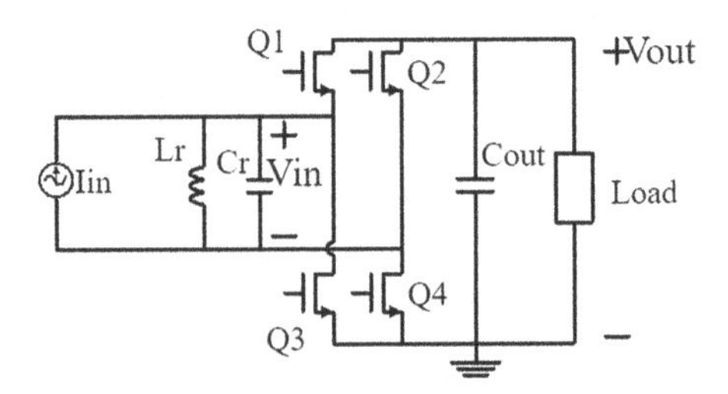

FIGURE 6.8 Synchronous and current-driven rectifier. (From Cochran, S. and Costinett, D., "Modeling a 6.78 MHz synchronous WPT rectifier with reduced THD," *2017 IEEE 18th Workshop on Control and Modeling for Power Electronics* (COMPEL), Stanford, CA, 1–8, 2017. With permission.)

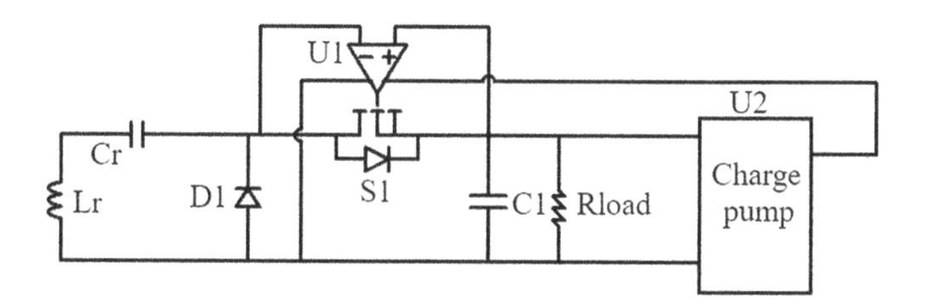

FIGURE 6.9 Self-driven synchronous rectifier. (From Koyama, T., Honjo, T., Ishihara, M., Umetani, K., and Hiraki, E., "Simple self-driven synchronous rectifier for resonant inductive coupling wireless power transfer," *2017 IEEE International Telecommunications Energy Conference* (INTELEC), Broadbeach, Australia, 363–368, 2017. With permission.)

An efficient active rectifier for WPT with a pulse width modulation (PWM)-controlled delay-compensated loop was proposed in Ref. [57] for getting 91% PCE. A 13.56-MHz wireless power transfer system with dual-output regulated active rectifier was developed in Ref. [81] for implantable medical devices. In Ref. [82], a comparative study of Class-E inverter and synchronous rectifier designs was made for wireless power applications operating in the 13.56, 27.12, and 40.68 MHz frequency bands. Ma et al. [54] proposed a ZVS active rectifier with adaptive on/off delay compensation for WPT systems using 0.18-μm CMOS process to get a PCE greater than 90% for a load resistor of 200 Ω. In Ref. [83], a synchronous rectifier to reduce the dissipated energy of the rectifier for almost arbitrary AC waveforms was presented. The approach can be used in almost any case and offers a simple method to improve the efficiency of WPT applications even when no dedicated integrated circuit can be used. An active single-phase rectifier (ASPR) with an auxiliary measurement coil (AMC) and a corresponding control method are proposed [84] for the MEPT in various conditions, such as reactance mismatch conditions and various load resistance conditions. In Ref. [85], an active rectifier with an output power of 1.37 mW was designed and simulated in a 130-nm CMOS technology for achieving an efficiency of 95.7%. In Ref. [86], the control of a 6.78-MHz GaN-based synchronous full-bridge rectifier was discussed, as shown in Figure 6.10. The rectifier was able to synchronize to the switching frequency of the carrier signal with controllable phase, allowing dynamic adjustment of the reactive load presented to the transmitter.

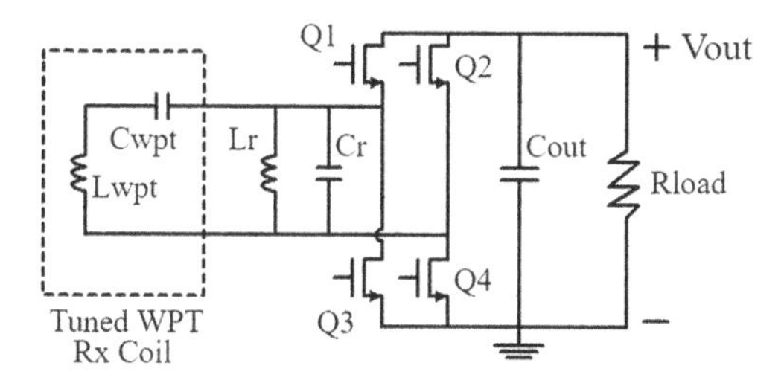

FIGURE 6.10 GaN-based rectifier. (From Cochran, S. and Costinett, D., "Frequency synchronization and control for a 6.78 MHz WPT active rectifier," *2018 IEEE 19th Workshop on Control and Modeling for Power Electronics* (COMPEL), Padua, 1–7, 2018. With permission.)

TABLE 6.1
Performance Comparison of Various Rectifiers

Ref.	Tech. (μm)	Chip Area (mm²)	Frequency (MHz)	V_{in} (V)	V_{out} (V)	P_{out} (mW)	PCE (%)	VCR
[74]	0.35	0.34	13.56	1.5–4	1.28–3.56, @Rl=1.8 kΩ 1.19–3.52, @Rl=500 Ω	24.8	82.2–90.1 @Rl=500 Ω	0.873–0.93, @ Rl=1.8 kΩ 0.79–0.89, @ Rl=500 Ω
[79]	0.18	3.45	6.78	7–20	3.45	8	94.2	
[81]	0.35		13.56	1.8–3.6		64.8	89.1–91.4 @Rl=500 Ω	0.92–0.946, @ Rl=2 kΩ 0.904–0.924, @ Rl=500 Ω
[77]	0.18		13.56	1.5–4.5	1.34–4.2	–	88.5–92.2 @Rl=500 Ω	0.89–0.93
[84]	0.065	1.44	13.56	1.3–2.5	1.24–2.44 @ Rl=500 Ω 1.2–2.39 @ Rl=100 Ω	248.1	88.5–91@ Rl=500 Ω 91.3–94.6 @Rl=100 Ω	0.948–0.977 @ Rl=500 Ω 0.917–0.952 @ Rl=100 Ω
[86]	0.13		13.56	1.2	1.09	44.93	81@ Rl=193 Ω	
[87]	0.18		6.78	5.26	5	1500	91.8	0.96
[88]	0.18	0.152	13.56	1.3–2	–	51.25	88.3–90.2@2 kΩ 92.3–94 @500 Ω Peak 95.3@200 Ω	0.967–0.97 @ 2 k, 90.7–91.5 @200
[89]	0.35	1.92	6.78	–	–	600	80.5	–
[101]	0.13	0.102	13.56	2.14–3.6	2–3.46	89	88–91.9 @ 3.6 Vac Rl=100 Ω–1 kΩ	0.86–0.964 @ 3.6 Vac, Rl=100–1 k
[116]	0.13	0.006	13.56	0.92–1.33	0.82–1.2	6.5	86.4–86.8 @Rl=200 Ω 85.1–85.7 @Rl=500 Ω	0.9–0.92 @ Rl=200 0.92–0.95 @Rl=500
[121]	0.18	0.117	13.56	1–2.5	–	34.1	85–94.1 @Rl=510 Ω	0.928–0.949 @Rl=510

In Ref. [87], a self-driven synchronous rectifier controller suitable for RIC-WPT was proposed. Zou et al. [88] proposed a secondary control method using an active rectifier for a double-sided LCC-tuned WPT to reduce the communication requirements of the system. In Ref. [89], a synchronous rectification-based phase-shift keying method for WPT systems was proposed. In Ref. [90], a synchronous rectifier phase-shift control method which provides the wireless EV charger control without any additional components was proposed. Zan et al. [91] proposed 27.12-MHz WPT rectifiers using current-mode Class-D (CMCD) converters to get the highest efficiency of 74.7% with 13.8 W output. Jensen et al. [92] reported a Class-DE full-bridge receiver circuit for WPT applications fulfilling the AirFuel standard. A 5-W rectifier uses 15-V GaN transistors to allow synchronous rectification and soft switching, thereby achieving a high efficiency. In Ref. [93], an active CMOS rectifier with new delay compensation technique for WPT system was proposed to get a maximum efficiency of 86.7% with 200 Ω load. Sinha et al. [94] developed a comprehensive methodology incorporated with an L-section matching network and the AVR rectifier for the design of large-air-gap capacitive WPT systems. In Ref. [95], a synchronous rectifier solution to improve the efficiency of receiving part of a WPT system was presented. Liu et al. [96] proposed an active Class-E rectifier for DC output voltage regulation in MHz WPT to get a rectifying efficiency of over 92%. In Ref. [97], an active rectifier structure based on adaptive delay time controller was described, which achieved a conversion efficiency of up to 94.1% with a static power consumption of not more than 230 μW. In Ref. [98], an adaptive delay time control (ADTC)-based CMOS active rectifier was proposed to get a peak 94.1% PCE at the output power of 10.63 mW. Song et al. [99] proposed an active Class-E rectifier for performing both high-frequency rectification and output voltage regulation.

6.5 RECONFIGURABLE RECTIFIERS

As the efficiency of the WPT system depends on the input power level and loading and coupling conditions, reconfigurable rectifiers that can adapt to the input level and loading and coupling conditions are desirable. A 1X/2X reconfigurable rectifier helps to achieve a higher system voltage gain and a higher system power transfer efficiency under different loading and coupling conditions [100]. By combining two circuit topologies for low-input-power and high-input-power and utilizing a depletion-mode field-effect transistor (FET) switch, the configuration of the rectifier can automatically be reconfigured to the more favorable circuit topology depending on the input power level [101]. A novel multiple-gain ×1-×2-×3 rectifier using a modular series–parallel architecture that allows a more conversion ratio and a wider output power range is enabled with a better impedance regulation to improve its efficiency by adjusting the receiver coil's impedance [102].

In Refs. [103,104], a power-efficient reconfigurable active voltage doubler/rectifier (VD/REC) was presented for robust WPT through inductive links over an extended range. In Ref. [101], a method to extend the operating input power range of a rectifier was proposed to get a conversion efficiency greater than 50% for

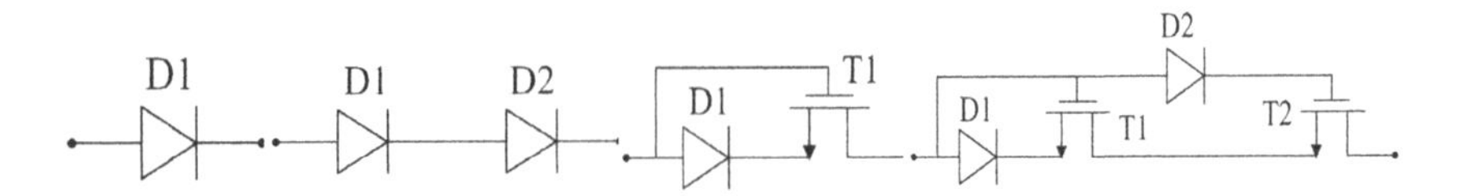

FIGURE 6.11 Different diode topologies. (From Liu, Z., Zhong, Z., and Guo, Y., *IEEE Micro. Wireless Comp. Lett.*, 26, 846–848, 2016. With permission.)

input power levels extending from –14 to 21 dBm. In Ref. [104], a reconfigurable diode topology (Figure 6.11) was presented to obtain independently tuned ultra-low threshold voltage and large breakdown voltage simultaneously. Nicoli et al. [102] proposed a novel multiple-gain ×1-×2-×3 reconfigurable rectifier using a modular series–parallel architecture.

In Ref. [105], a simple and effective technique was presented to reduce the limit cycle oscillations (LCO) in the reconfigurable rectifier for wireless power receiver using 0.18-μm CMOS technology. In Ref. [106], a 1X/2X reconfigurable rectifier with compact structure and consistent input capacitance under different modes was presented for WPT in implantable medical devices (IMDs). In Ref. [107], an efficient and reconfigurable rectifier circuit with the capability of automatically switching from low-power to high-power operation mode was presented. In Ref. [108], a reconfigurable rectifier was presented to harvest energy with net-zero energy consumption, i.e., without draining power from battery, in a wide input power range. Liu et al. [109] presented a dual-band WPT architecture with novel transmitter and receiver topologies to achieve high performance at both 100 kHz and 13.56 MHz.

6.6 RESONANT REGULATING RECTIFIERS

WPTs can be classified into two types: inductively coupled power transfer (IPT) and resonant WPT (RWPT) systems. IPT has higher efficiency, but requires very short distances and precise alignment between the transmitter and the receiver coils, whereas RWPT uses loosely coupled coils to achieve better spatial freedom [110]. For IMDs, the coil coupling varies with the distance and the alignment between the coils. Both coupling and load variations make the output voltage of the rectifier unsteady. If there is no power control, the worst-case scenario has to be designed for when the coupling distance is large, and a high enough transmission power should be used to maintain the output voltage higher than the minimum required value. However, for cases with better coupling conditions, the output voltage becomes much higher than the required value [111]. Hence, the output voltage regulation is needed for most of the converter circuits for IMDs and wireless charging devices. This can be achieved by employing a regulating technique at the receiver end for better response and precise voltage condition [111]. The resonant regulating rectifier (3R) and reconfigurable resonant regulating (R^3) rectifier have ability to solve these issues. An R^3 rectifier reconfigures the output stage to upkeep the output voltage of the rectifier when the coupling conditions change. Choi et al. [110] presented new receiver circuits for the RWPT, using resonant regulating rectifier (3R), with a power transfer capability of more than 5 W and an independent regulating ability.

In Ref. [112], a 13.56-MHz WPT system with a 1X/2X reconfigurable resonant regulating (R^3) rectifier and wireless power control for biomedical implants was presented to get 102 mW maximum received power and 92.6% receiver efficiency. A 13.56-MHz WPT system was proposed in Ref. [111] by employing a reconfigurable resonant regulating rectifier for V_{out} regulation in coupling and loading variation conditions.

In Ref. [113], a reconfigurable resonant regulating rectifier with primary equalization for extended coupling and loading range in bio-implant WPT was proposed. Kim et al. [5] proposed an integrated 3R for low-power inductive power telemetry in implantable devices. An adaptive buck–boost resonant regulating rectifier (B2R3) with an integrated on-chip coil and low-loss H-tree power/signal distribution was presented by Kim et al. [114] for efficient and robust WPT over a wide range of input and load conditions. Cheng et al. [115] presented a 6.78-MHz 6-W wireless power receiver having a three-level 1X/½ X/0X reconfigurable resonant regulating rectifier with PWM control for mode switching. Cheng et al. [116] proposed a 6.78-MHz wireless power receiver using a three-mode reconfigurable resonant regulating rectifier for resonant WPT. Cheng et al. [117] proposed a wireless power receiver with the aim of improving the performance and reducing the cost of a three-level reconfigurable resonant regulating (R^3) rectifier and obtained a peak efficiency of 92.2%. Kim et al. [118] proposed a fully integrated resonant regulating rectifier (IR3) that performs voltage and power conversion from an integrated 3×3 mm^2 on-chip coil to loads in which no external components are required for operation. In Ref. [119], a new type of fully integrated magnetic resonance wireless power receiver IC was presented, which reduced the received power and regulated output voltage by shifting the resonance frequency. In Ref. [120], a 13.56-MHz wireless power receiver designed with standard 0.35-μm CMOS process used for implantable medical devices (IMDs) was presented for getting a maximum output power of 103.5 mW and an efficiency of 91.7%. In Ref. [121], a 200 MHz WPT system with resonant reconfigurable rectifier is presented to establish data link using backscattering communication for biomedical applications. A 13.56-MHz regulated dual-output active rectifier was proposed for implantable medical devices (IMDs). Ghotbi et al. [122] proposed a resonant regulating rectifier (R^3) topology based on a switchable quadrupler in order to regulate the output voltage on 20 V in a wide coupling range without making use of DC–DC converters or linear regulators. Cheng et al. [123] proposed a simplified PWM controller for the wireless power receiver using a three-mode reconfigurable resonant regulating (R^3) rectifier. In Ref. [124], a fully integrated wireless-power-receiver-on-chip (WiPRoC) with an adaptive buck–boost resonant regulating rectifier (B2R3) and a low-loss H-tree power/signal distribution network was presented. In Ref. [125], a small low-cost self-regulated rectifier was proposed for parallel resonant receivers (RXs) in WPT system to maintain the individual rectifier voltages of each RX to be constant. In Ref. [126], a single-switch-regulated WPT receiver in which one active switch was used without any diode was presented.

REFERENCES

1. Y. Sun, C. Jeong, S. Han, and S. Lee, "A high speed comparator based active rectifier for wireless power transfer systems," *2011 IEEE MTT-S International Microwave Workshop Series on Intelligent Radio for Future Personal Terminals*, Daejeon, pp. 1–2, 2011.
2. D. Huwig and P. Wambsganß, "Digitally controlled synchronous bridge-rectifier for wireless power receivers," *2013 Twenty-Eighth Annual IEEE Applied Power Electronics Conference and Exposition* (APEC), Long Beach, CA, pp. 2598–2603, 2013.
3. H. Cha, W. Park, and M. Je, "A CMOS rectifier with a cross-coupled latched comparator for wireless power transfer in biomedical applications," *IEEE Transactions on Circuits and Systems II: Express Briefs*, vol. 59, no. 7, pp. 409–413, July 2012.
4. K. Inoue, T. Nagashima, X. Wei, and H. Sekiya, "Design of high-efficiency inductive-coupled wireless power transfer system with Class-DE transmitter and Class-E rectifier," *IECON 2013 - 39th Annual Conference of the IEEE Industrial Electronics Society*, Vienna, pp. 613–618, 2013.
5. P.C.K. Luk and S. Aldhaher, "Analysis and design of a Class D rectifier for a Class E driven wireless power transfer system," *IEEE Energy Conversion Congress and Exposition* (ECCE), Pittsburgh, PA, pp. 851–857, 2014.
6. M. Liu, M. Fu, and C. Ma, "A compact Class E rectifier for megahertz wireless power transfer," *2015 IEEE PELS Workshop on Emerging Technologies: Wireless Power* (2015 WoW), Daejeon, pp. 1–5, 2015.
7. T. Nagashima, X. Wei, and H. Sekiya, "Analytical design procedure for resonant inductively coupled wireless power transfer system with Class-DE inverter and Class-E rectifier," *2014 IEEE Asia Pacific Conference on Circuits and Systems* (APCCAS), Ishigaki, pp. 288–291, 2014.
8. G. Kkelis, D.C. Yates, and P.D. Mitcheson, "Hybrid Class-E low dv/dt rectifier for high frequency inductive power transfer," *2016 IEEE Wireless Power Transfer Conference* (WPTC), pp. 1–4, 2016.
9. G. Kkelis, D.C. Yates, and P.D. Mitcheson, "Comparison of current driven Class-D and Class-E half-wave rectifiers for 6.78 MHz high power IPT applications," *IEEE Wireless Power Transfer Conference* (WPTC), Boulder, CO, pp. 1–4, 2015.
10. T. Nagashima, X. Wei, E. Bou, E. Alarc6n, and H. Sekiya, "Analytical design for resonant inductive coupling wireless power transfer system with Class-E inverter and Class-DE rectifier," *2015 IEEE International Symposium on Circuits and Systems* (ISCAS), Lisbon, pp. 686–689, 2015.
11. M.K. Kazimierczuk and J. Jozwik, "Resonant DC/DC converter with Class-E inverter and Class-E rectifier," *IEEE Transactions on Industrial Electronics*, vol. 36, no. 4, pp. 468–478, November 1989.
12. M. Bojarski, E. Asa, and D. Czarkowski, "Three-phase resonant inverter for wireless power transfer," *2015 IEEE Wireless Power Transfer Conference* (WPTC), Boulder, CO, pp. 1–4, 2015.
13. M. Liu, M. Fu, and C. Ma, "Parameter design for a 6.78-MHz wireless power transfer system based on analytical derivation of Class E current-driven rectifier," *IEEE Transactions on Power Electronics*, vol. 31, no. 6, pp. 4280–4291, June 2016.
14. S. Aldhaher, D.C. Yates, and P.D. Mitcheson, "Design and development of a Class EF2 inverter and rectifier for multimegahertz wireless power transfer systems," *IEEE Transactions on Power Electronics*, vol. 31, no. 12, pp. 8138–8150, December 2016.
15. L. Rizo, D. Vegas, M.N. Ruiz, R. Marante, L. Cabria, and J.A. García, "Class-E amplifier and rectifier for a wireless link with secure signal and simultaneous power transmission," *IEEE Wireless Power Transfer Conference* (WPTC), Aveiro, pp. 1–3, 2016.

16. M. Liu, Y. Qiao, and C. Ma, "Robust optimization for a 6.78-MHz wireless power transfer system with Class E rectifier," *2016 IEEE PELS Workshop on Emerging Technologies: Wireless Power Transfer* (WoW), Knoxville, TN, pp. 88–94, 2016.
17. M. Liu, M. Fu, and C. Ma, "Low-harmonic-contents and high-efficiency Class E full-wave current-driven rectifier for megahertz wireless power transfer systems," *IEEE Transactions on Power Electronics*, vol. 32, no. 2, pp. 1198–1209, February 2017.
18. T. Inaba, H. Koizumi, and H. Sekiya, "Design of wireless power transfer system with Class E inverter and half-bridge Class DE rectifier at any fixed coupling coefficient," *2017 IEEE 3rd Int. Future Energy Electronics Conference and ECCE Asia* (IFEEC 2017-ECCE Asia), Kaohsiung, pp. 185–189, 2017.
19. X. Fu, M. Liu, Z. Tang, and C. Ma, "Design procedure of a Class E2 DC-DC converter for megahertz wireless power transfer based on a compact Class E current-driven rectifier," *2017 IEEE 26th International Symposium on Industrial Electronics* (ISIE), Edinburgh, pp. 694–699, 2017.
20. X. Zhang, X. Zhang, Y. Yao, H. Yang, Y. Wang, and D. Xu, "High-efficiency magnetic coupling resonant wireless power transfer system with Class-E amplifier and Class-E rectifier," *2017 IEEE Transportation Electrification Conf. and Expo, Asia-Pacific* (ITEC Asia-Pacific), Harbin, pp. 1–5, 2017.
21. M. Liu, S. Liu, and C. Ma, "A high-efficiency/output power and low-noise megahertz wireless power transfer system over a wide range of mutual inductance," *IEEE Transactions on Microwave Theory and Techniques*, vol. 65, no. 11, pp. 4317–4325, November 2017.
22. S. Liu, M. Liu, S. Han, X. Zhu, and C. Ma, "Tunable Class E^2 DC-DC converter with high efficiency and stable output power for 6.78-MHz wireless power transfer," *IEEE Transactions on Power Electronics*, vol. 33, no. 8, pp. 6877–6886, August 2018.
23. S. Mita and H. Sekiya, "Analysis and design of wireless power transfer system with asymmetrical duty-cycle controlled Class-D ZVS inverter," *2018 IEEE International Symposium on Circuits and Systems* (ISCAS), Florence, pp. 1–5, 2018.
24. F.C. Domingos, S.V.D.C. de Freitas, and P. Mousavi, "Capacitive power transfer based on compensation circuit for Class E resonant full-wave rectifier," *2018 IEEE Wireless Power Transfer Conference* (WPTC), Montreal, QC, Canada, pp. 1–4, 2018.
25. S. Liu and C. Ma, "Low-harmonic-distortion and high-efficiency Class E2 DC-DC converter for 6.78 MHz WPT," *2018 IEEE International Conference on Industrial Electronics for Sustainable Energy Systems* (IESES), Hamilton, pp. 421–425, 2018.
26. S. Yang, K. Chen, Y. Lin, S. Lin, and T. Tsai, "A temperature compensated 61-W Class-E soft-switching GaN-based active diode rectifier for wireless power transfer applications," *IEEE Solid-State Circuits Letters*, vol. 2, no. 9, pp. 203–206, September 2019.
27. X. Zhu, K. Jin, and Y. Chen, "Adaptive impedance matching of rectifier for a 100 MHz microwave power transfer system," in *IET Power Electronics*, vol. 12, no. 9, pp. 2353–2360, 2019.
28. Y. Zhang and M.D. Rooij, "Rectifier topology comparison in 6.78 MHz highly resonant wireless power systems," *2019 IEEE Applied Power Electronics Conference and Exposition* (APEC), Anaheim, CA, pp. 671–677, 2019.
29. A. Bati, P.C.K. Luk, S. Aldhaher, C.H. See, R. A. Abd-Alhameed, and P.S. Excell, "Efficiency improvement of a Class E2 converter for low power inductive links," *2019 26th International Workshop on Electric Drives: Improvement in Efficiency of Electric Drives* (IWED), Moscow, Russia, pp. 1–6, 2019.
30. F. Zhao, D. Inserra, G. Wen, J. Li, and Y. Huang, "A high-efficiency inverse Class-F microwave rectifier for wireless power transmission," *IEEE Microwave and Wireless Components Letters*, vol. 29, no. 11, pp. 725–728, November 2019.

31. M. Zargham and P.G. Gulak, "High-efficiency CMOS rectifier for fully integrated mW wireless power transfer," *2012 IEEE International Symposium on Circuits and Systems* (ISCAS), pp. 2869–2872, 2012.
32. S. Hashemi, M. Sawan, and Y. Savaria, "Fully-integrated low-voltage high-efficiency CMOS rectifier for wirelessly powered devices," *2009 Joint IEEE North-East Workshop on Circuits and Systems and TAISA Conference*, Toulouse, pp. 1–4, 2009.
33. S. Guo and H. Lee, "An efficiency-enhanced CMOS rectifier with unbalanced-biased comparators for transcutaneous-powered high-current implants," *IEEE Journal of Solid-State Circuits*, vol. 44, no. 6, pp. 1796–1804, June 2009.
34. G. Bawa, A.Q. Huang, and M. Ghovanloo, "An efficient 13.56 MHz active back-telemetry rectifier in standard CMOS technology," *Proceedings of 2010 IEEE International Symposium on Circuits and Systems*, Paris, pp. 1201–1204, 2010.
35. M. Karimi and H. Nabovati, "Design of a high efficient fully integrated CMOS rectifier using bootstrapped technique for sub-micron and wirelessly powered applications," *18th IEEE International Conference on Electronics, Circuits, and Systems*, Beirut, pp. 133–136, 2011.
36. S.S. Hashemi, M. Sawan, and Y. Savaria, "A high-efficiency low-voltage CMOS rectifier for harvesting energy in implantable devices," *IEEE Trans on Biomedical Circuits and Systems*, vol. 6, no. 4, pp. 326–335, August 2012.
37. B.W. Ha and C.S. Cho, "Design of rectifier with comparator using unbalanced body biasing for wireless power transfer," *2013 International SoC Design Conference* (ISOCC), Busan, pp. 356–359, 2013.
38. A. Mansano, S. Bagga, and W. Serdijn, "A high efficiency orthogonally switching passive charge pump rectifier for energy harvesters," *IEEE Transactions on Circuits and Systems I: Regular Papers*, vol. 60, no. 7, pp. 1959–1966, July 2013.
39. Q. Li, J. Wang, and Y. Inoue, "A high efficiency CMOS rectifier with ON-OFF response compensation for wireless power transfer in biomedical applications," *2014 International Symposium on Integrated Circuits* (ISIC), Singapore, pp. 91–94, 2014.
40. R. Ramzan and F. Zafar, "High-efficiency fully CMOS VCO rectifier for microwatt resonant wireless power transfer," *IEEE Transactions on Circuits and Systems II: Express Briefs*, vol. 62, pp. 134–138, 2015.
41. M. Mahmoud, "Efficiency improvement of differential drive rectifier for wireless power transfer applications," *2016 7th International Conference on Intelligent Systems, Modelling and Simulation* (ISMS), Bangkok, pp. 435–439, 2016.
42. Deylamani, M.J., Abdi, F., and Amiri, P, "A full-wave CMOS rectifier with high-speed comparators for implantable medical devices," *Journal of Circuits, Systems and Computers*, vol. 28, no. 11, pp. 1950178, 2019.
43. N.T. Tasneem, S.R. Suri, and I. Mahbub, "A low-power CMOS voltage boosting rectifier for wireless power transfer applications," *2018 Texas Symposium on Wireless and Microwave Circuits and Systems* (WMCS), Waco, TX, pp. 1–4, 2018.
44. Z. Ehsanbakhsh and M. Jalali, "Optimum design of RF-to-DC energy conversion circuits for wireless powered applications," *Electrical Engineering (ICEE), Iranian Conference on*, Mashhad, pp. 325–329, 2018.
45. S. Cochran, F. Quaiyum, A. Fathy, D. Costinett, and S. Yang, "A GaN-based synchronous rectifier for WPT receivers with reduced THD," *2016 IEEE PELS Workshop on Emerging Technologies: Wireless Power Transfer* (WoW), Knoxville, TN, pp. 81–87, 2016.
46. J.R. de Castilho Louzada, L.B. Zoccal, R.L. Moreno, and T.C. Pimenta, "A high efficiency 0.13μm CMOS full wave active rectifier with comparators for implanted medical devices," *Advances in Science, Technology and Engineering Systems Journal*, vol. 2, no. 3, pp. 1019–1025, 2017.

47. D. Fu, Y. Liu, F.C. Lee, and M. Xu, "A novel driving scheme for synchronous rectifiers in LLC resonant converters," *IEEE Transactions on Power Electronics*, vol. 24, no. 5, pp. 1321–1329, 2009.
48. Y. Xiang, Y. Wang, and C.R. Shi, "A 13.56 MHz active rectifier with self-switching comparator for wireless power transfer systems," *2018 International SoC Design Conference* (ISOCC), Daegu, Korea (South), pp. 54–55, 2018.
49. J. Fuh, S. Hsieh, F. Yang, and P. Chen, "A 13.56MHz power-efficient active rectifier with digital offset compensation for implantable medical devices," *2016 IEEE Wireless Power Transfer Conference* (WPTC), Aveiro, pp. 1–3, 2016.
50. Y. Lu and W. Ki, "A 13.56 MHz CMOS active rectifier with switched-offset and compensated biasing for biomedical wireless power transfer systems," *IEEE Transactions on Biomedical Circuits and Systems*, vol. 8, no. 3, pp. 334–344, June 2014.
51. L. Cheng, W. Ki, Y. Lu, and T. Yim, "Adaptive on/off delay-compensated active rectifiers for wireless power transfer systems," *IEEE Journal of Solid-State Circuits*, vol. 51, pp. 712–723, 2016.
52. L. Cheng, W. Ki, and T. Yim, "A 13.56 MHz on/off delay-compensated fully-integrated active rectifier for biomedical wireless power transfer systems," *2017 22nd Asia and South Pacific Design Automation Conference* (ASP-DAC), Chiba, pp. 31–32, 2017.
53. F. Mao, Y. Lu, U. Seng-Pan, and R.P. Martins, "A 6.78 MHz active voltage doubler with near-optimal on/off delay compensation for wireless power transfer systems," *2018 International Symposium on VLSI Design, Automation and Test* (VLSI-DAT), Hsinchu, pp. 1–4, 2018.
54. Y. Ma, K. Cui, Z. Ye, S. Fang, and X. Fan, "A ZVS active rectifier with adaptive on/off delay compensation for WPT systems," *2018 IEEE Asia Pacific Conference on Circuits and Systems* (APCCAS), Chengdu, pp. 167–170, 2018.
55. X. Bai, Z. Kong, and L. Siek, "A high-efficiency 6.78-MHz full active rectifier with adaptive time delay control for wireless power transmission," *IEEE Transactions on Very Large Scale Integration (VLSI) Systems*, vol. 25, no. 4, pp. 1297–1306, April 2017.
56. Z. Xue, D. Li, W. Gou, L. Zhang, S. Fan, and L. Geng, "A delay time controlled active rectifier with 95.3% peak efficiency for wireless power transmission systems," *2017 IEEE International Symposium on Circuits and Systems* (ISCAS), Baltimore, MD, pp. 1–4, 2017.
57. Y. Ma, K. Cui, Z. Ye, S. Fang, and X. Fan, "A 13.56MHz integrated CMOS active rectifier with adaptive delay-compensated loop for wireless power transfer systems," *2018 14th IEEE International Conference on Solid-State and Integrated Circuit Technology* (ICSICT), Qingdao, pp. 1–3, 2018.
58. H. Cheng, C.A. Gong, and S. Kao, "A 13.56 MHz CMOS high-efficiency active rectifier with dynamically controllable comparator for biomedical wireless power transfer systems," *IEEE Access*, vol. 6, pp. 49979–49989, 2018.
59. M.S. Chinthavali, O.C. Onar, J.M. Miller, and L. Tang, "Single-phase active boost rectifier with power factor correction for wireless power transfer applications," *2013 IEEE Energy Conversion Congress and Exposition*, Denver, CO, pp. 3258–3265, 2013.
60. L. Tang, M. Chinthavali, O.C. Onar, S. Campbell, and J.M. Miller, "SiC MOSFET based single phase active boost rectifier with power factor correction for wireless power transfer applications," *2014 IEEE Applied Power Electronics Conference and Exposition-APEC 2014*, Fort Worth, TX, pp. 1669–1675, 2014.
61. K. Colak, E. Asa, M.Bojarski, D. Czarkowski, and O.C. Onar, "A novel phase-shift control of semi bridgeless active rectifier for wireless power transfer," *IEEE Transactions on Power Electronics*, vol. 30, no. 11, pp. 6288–6297, November 2015.

62. A. Berger, M. Agostinelli, S. Vesti, J. A. Oliver, J. A. Cobos, and M. Huemer, "Phase-shift and amplitude control for an active rectifier to maximize the efficiency and extracted power of a Wireless Power Transfer system," *2015 IEEE Applied Power Electronics Conference and Exposition* (APEC), Charlotte, NC, pp. 1620–1624, 2015.
63. J. Fuh, S. Hsieh, F. Yang, and P. Chen, "A 13.56 MHz power-efficient active rectifier with digital offset compensation for implantable medical devices," *2016 IEEE Wireless Power Transfer Conference* (WPTC), Aveiro, pp. 1–3, 2016.
64. Y.-J. Park, S. Oh, S. Kim, S. Cho, M. Kim, J.H. Park, D. Lee, H. Kim, and K.Y. Lee, "A design of inductive coupling wireless power receiver with high efficiency Active Rectifier and multi feedback LDO regulator," *2016 IEEE Wireless Power Transfer Conference* (WPTC), Aveiro, pp. 1–4, 2016.
65. Y.-J. Park, H.G. Park, J. Lee, S.J. Oh, J.H. Jang, S.Y. Kim, Y.G. Pu, K.C. Hwang, Y. Yang, M. Seo, and K.Y. Lee, "A design of wide input range triple-mode active rectifier with peak efficiency of 94.2 % and maximum output power of 8 W for wireless power receiver in 0.18 lM BCD," *Analog Integrated Circuits and Signal Processing*, vol. 86, pp. 255–265, 2016.
66. L. Cheng, W. Ki, Y. Lu, and T. Yim, "Adaptive On/Off Delay-Compensated Active Rectifiers for Wireless Power Transfer Systems," in *IEEE Journal of Solid-State Circuits*, vol. 51, 712–723, 2016.
67. H. Kim, S. Jeong, D. Kim, J. Kim, Y. Kim, and I. Kim, "Selective harmonic elimination method of radiation noise from automotive wireless power transfer system using active rectifier," *2016 IEEE 25th Conference on Electrical Performance of Electronic Packaging and Systems* (EPEPS), San Diego, CA, pp. 161–164, 2016.
68. B. Jang, S. Oh, Y. Park, and K. Y. Lee, "A high efficiency active rectifier with zero current sensing for loosely-coupled wireless power transfer systems," *2016 IEEE International Conference on Consumer Electronics-Asia* (ICCE-Asia), Seoul, pp. 1–2, 2016.
69. C. Huang, T. Kawajiri, and H. Ishikuro, "A near-optimum 13.56 MHz CMOS active rectifier with circuit-delay real-time calibrations for high-current biomedical implants," *IEEE Journal of Solid-State Circuits*, vol. 51, no. 8, pp. 1797–1809, August 2016.
70. J.R.C. Louzada, L.B. Zoccal, R.L. Moreno, and T.C. Pimenta, "A 0.13μm CMOS full wave active rectifier with comparators for implanted medical devices," *2016 International Symposium on Integrated Circuits* (ISIC), Singapore, pp. 1–4, 2016.
71. Y. Lu, M. Huang, L. Cheng, W. Ki, U. Seng-Pan, and R.P. Martins, "A dual-output wireless power transfer system with active rectifier and three-level operation," *IEEE Transactions on Power Electronics*, vol. 32, no. 2, pp. 927–930, February 2017.
72. M. Kerber, B. Offord, and A. Phipps, "Design considerations for an active rectifier circuit for bidirectional wireless power transfer," *2017 IEEE Wireless Power Transfer Conference* (WPTC), Taipei, pp. 1–4, 2017.
73. G. Kkelis, S. Aldhaher, J.M. Arteaga, D.C. Yates, and P.D. Mitcheson, "Hybrid Class-E synchronous rectifier for wireless powering of quadcopters," *2017 IEEE Wireless Power Transfer Conference* (WPTC), Taipei, pp. 1–4, 2017.
74. K. Hata, T. Imura, and Y. Hori, "Maximum efficiency control of wireless power transfer systems with half active rectifier based on primary current measurement," *2017 IEEE 3rd International Future Energy Electronics Conference and ECCE Asia* (IFEEC 2017 - ECCE Asia), Kaohsiung, pp. 1–6, 2017.
75. S. Cochran and D. Costinett, "Modeling a 6.78 MHz synchronous WPT rectifier with reduced THD," *2017 IEEE 18th Workshop on Control and Modeling for Power Electronics* (COMPEL), Stanford, CA, pp. 1–8, 2017.
76. T. Koyama, T. Honjo, M. Ishihara, K. Umetani, and E. Hiraki, "Simple self-driven synchronous rectifier for resonant inductive coupling wireless power transfer," *2017 IEEE International Telecommunications Energy Conference* (INTELEC), Broadbeach, Australia, pp. 363–368, 2017.

77. F. Yang, S. Hsieh, and P. Chen, "A 13.56 MHz pulse-width modulation active rectifier for implantable medical devices," *2017 IEEE Wireless Power Transfer Conference* (WPTC), Taipei, pp. 1–4, 2017.
78. S. Aldhaher, D.C. Yates, and P.D. Mitcheson, "13.56MHz 50W load-independent synchronous Class E rectifier using GaN devices for space-constrained applications," *2018 IEEE Wireless Power Transfer Conference* (WPTC), Montreal, QC, Canada, pp. 1–4, 2018.
79. R. Gallichan, D.M. Budgett, and D. McCormick, "600 mW active rectifier with shorting-control for wirelessly powered medical implants," *2018 IEEE Biomedical Circuits and Systems Conference* (BioCAS), Cleveland, OH, pp. 1–4, 2018.
80. H. Cheng, C.A. Gong, and S. Kao, "A 13.56 MHz CMOS High-Efficiency Active Rectifier with Dynamically Controllable Comparator for Biomedical Wireless Power Transfer Systems," in *IEEE Access*, vol. 6, pp. 49979–49989, 2018.
81. F. Yang, J. Fuh, and P. Chen, "A 13.5 MHz wireless power transfer system with dual-output regulated active rectifier for implantable medical devices," *2018 IEEE 61st International Midwest Symposium on Circuits and Systems* (MWSCAS), Windsor, ON, Canada, pp. 440–443, 2018.
82. A. Clements, V. Vishnoi, S. Dehghani, and T. Johnson, "A comparison of GaN Class E inverter and synchronous rectifier designs for 13.56 MHz, 27.12 MHz and 40.68 MHz ISM bands," *2018 IEEE Wireless Power Transfer Conference* (WPTC), Montreal, QC, Canada, pp. 1–4, 2018.
83. S. Mauch, H. Reichle, and D. Benyoucef, "Synchronous rectifier for high-power wireless transfer applications," *IECON 2018-44th Annual Conference of the IEEE Industrial Electronics Society*, Washington, DC, pp. 288–293, 2018.
84. R. Mai, Y. Liu, Y. Li, P. Yue, G. Cao, and Z. He, "An active-rectifier-based maximum efficiency tracking method using an additional measurement coil for wireless power transfer," *IEEE Transactions on Power Electronics*, vol. 33, no. 1, pp. 716–728, January 2018.
85. P. Perez-Nicoli, F. Veirano, and F. Silveira, "Comparator with self controlled delay for active rectifiers in inductive powering," *2018 IEEE Wireless Power Transfer Conference* (WPTC), Montreal, QC, Canada, pp. 1–4, 2018.
86. S. Cochran and D. Costinett, "Frequency synchronization and control for a 6.78 MHz WPT active rectifier," *2018 IEEE 19th Workshop on Control and Modeling for Power Electronics* (COMPEL), Padua, pp. 1–7, 2018.
87. A. Konishi, K. Umetani, and E. Hiraki, "High-frequency self-driven synchronous rectifier controller for WPT systems," *2018 International Power Electronics Conference* (IPEC-Niigata 2018 -ECCE Asia), Niigata, pp. 1602–1609, 2018.
88. S. Zou, O.C. Onar, V. Galigekere, J. Pries, G. Su, and A. Khaligh, "Secondary active rectifier control scheme for a wireless power transfer system with double-sided LCC compensation topology," *IECON 2018-44th Annual Conference of the IEEE Industrial Electronics Society*, Washington, DC, pp. 2145–2150, 2018.
89. H. Li, S. Chen, J. Fang, and Y. Tang, "Synchronous rectification-based phase shift keying communication for wireless power transfer systems," *2018 IEEE 4th Southern Power Electronics Conference* (SPEC), Singapore, Singapore, pp. 1–4, 2018.
90. T. Takahashi, H. Omori, T. Morizane, and N. Kimura, "A new type of high-power wireless power transfer with a superimposed communication by a phase-shifted synchronous-rectifier," *2018 IEEE International Power Electronics and Application Conference and Exposition* (PEAC), pp. 1–5, 2018.
91. X. Zan and A. Avestruz, "Performance comparisons of synchronous and uncontrolled rectifiers for 27.12 MHz wireless power transfer using CMCD converters," *2018 IEEE Energy Conversion Congress and Exposition* (ECCE), Portland, OR, pp. 2448–2455, 2018.

92. C.H.K. Jensen, F.M. Spliid, J.C. Hertel, Y. Nour, T. Zsurzsan, and A. Knott, "Resonant full-bridge synchronous rectifier utilizing 15 V GaN transistors for wireless power transfer applications following AirFuel standard operating at 6.78 MHz," *2018 IEEE Applied Power Electronics Conference and Exposition* (APEC), San Antonio, TX, pp. 3131–3137, 2018.
93. S. Shahsavari and M. Saberi, "A power-efficient CMOS active rectifier with circuit delay compensation for wireless power transfer systems," *Circuits, Systems, and Signal Processing*, vol. 38, pp. 947–966, 2019.
94. S. Sinha, B. Regensburger, A. Kumar, and K.K. Afridi, "A multi-MHz large air-gap capacitive wireless power transfer system utilizing an active variable reactance rectifier suitable for dynamic electric vehicle charging," *2019 IEEE Energy Conversion Congress and Exposition* (ECCE), Baltimore, MD, pp. 5726–5732, 2019.
95. K. Krestovnikov, E. Cherskikh, and N. Pavliuk, "Concept of a synchronous rectifier for wireless power transfer system," *IEEE EUROCON 2019-18th International Conference on Smart Technologies*, Novi Sad, Serbia, pp. 1–5, 2019.
96. M. Liu, J. Song, and C. Ma, "Active Class E rectifier for DC output voltage regulation in megahertz wireless power transfer systems," *IEEE Transactions on Industrial Electronics*, vol. 67, pp. 3618–3628, 2019.
97. L. Geng, Z. Xue, S. Fan, D. Li, and B. Zhang, "Active rectifiers in wireless power transmission systems," *2019 International Conference on IC Design and Technology* (ICICDT), Suzhou, China, pp. 1–3, 2019.
98. Z. Xue, S. Fan, D. Li, L. Zhang, W. Gou, and L. Geng, "A 13.56 MHz, 94.1% peak efficiency CMOS active rectifier with adaptive delay time control for wireless power transmission systems," *IEEE Journal of Solid-State Circuits*, vol. 54, no. 6, pp. 1744–1754, June 2019.
99. J. Song, M. Liu, and C. Ma, "Active Class E rectifier with controlled output voltage for megahertz wireless power transfer," *2019 IEEE 13th International Conference on Compatibility, Power Electronics and Power Engineering* (CPE-POWERENG), Sonderborg, Denmark, pp. 1–5, 2019.
100. X. Li, C. Tsui, and W. Ki, "Power management analysis of inductively-powered implants with 1X/2X reconfigurable rectifier," *IEEE Transactions on Circuits and Systems I: Regular Papers*, vol. 62, no. 3, pp. 617–624, March 2015.
101. H. Sun, Z. Zhong, and Y. Guo, "An adaptive reconfigurable rectifier for wireless power transmission," *IEEE Microwave and Wireless Components Letters*, vol. 23, no. 9, pp. 492–494, September 2013.
102. P.P. Nicoli and F. Silveira, "Reconfigurable multiple-gain active-rectifier for maximum efficiency point tracking in WPT," *2017 IEEE 8th Latin American Symposium on Circuits & Systems* (LASCAS), Bariloche, pp. 1–4, 2017.
103. H. Lee and M. Ghovanloo, "An adaptive reconfigurable active voltage doubler/rectifier for extended-range inductive power transmission," *2012 IEEE International Solid-State Circuits Conference*, San Francisco, CA, pp. 286–288, 2012.
104. Z. Liu, Z. Zhong, and Y. Guo, "A reconfigurable diode topology for wireless power transfer with a wide power range," *IEEE Microwave and Wireless Components Letters*, vol. 26, pp. 846–848, 2016.
105. X. Tang, J. Zeng, Y. Zheng, K.N. Leung, and Z. Wang, "Limit-cycle oscillation reduction in high-efficiency wireless power receiver," *Electronics Letters*, vol. 53, pp. 1152–1154, 2017.
106. G. Zhu, S. Mai, X. Tang, and Z. Wang, "Simplified reconfigurable rectifier with consistent input capacitance for wireless power transfer," *2018 IEEE International Conference on Electron Devices and Solid State Circuits* (EDSSC), Shenzhen, pp. 1–2, 2018.

107. T. Ngo, A. Huang, and Y. Guo, "Analysis and design of a reconfigurable rectifier circuit for wireless power transfer," *IEEE Transactions on Industrial Electronics*, vol. 66, no. 9, pp. 7089–7098, September 2019.
108. U. Guler, Y. Jia, and M. Ghovanloo, "A reconfigurable passive RF-to-DC converter for wireless IoT applications," *IEEE Transactions on Circuits and Systems II: Express Briefs*, vol. 66, pp. 1800–1804, 2019.
109. M. Liu and M. Chen, "Dual-band wireless power transfer with reactance steering network and reconfigurable receivers," *IEEE Transactions on Power Electronics*, vol. 35, pp. 496–507, January 2020.
110. J. Choi, S. Yeo, S. Park, J. Lee, and G. Cho, "Resonant regulating rectifiers (3R) operating for 6.78 MHz resonant wireless power transfer (RWPT)," *IEEE Journal of Solid-State Circuits*, vol. 48, no. 12, pp. 2989–3001, December 2013.
111. Li, X., Tsui, C.Y., and Ki, W.H. (2015). "A 13.56 MHz wireless power transfer system with reconfigurable resonant regulating rectifier and wireless power control for implantable medical devices," *IEEE Journal of Solid-State Circuits*, vol. 50, no. 4, pp. 978–989, April 2015.
112. X. Li, C. Ying Tsui, and W.H. Ki, "A 13.56 MHz wireless power transfer system with reconfigurable resonant regulating rectifier and wireless power control for implantable medical devices," *2014 Symposium on VLSI Circuits Digest of Technical Papers*, Honolulu, HI, pp. 1–2, 2014.
113. X. Li, X. Meng, C. Tsui, and W. Ki, "Reconfigurable resonant regulating rectifier with primary equalization for extended coupling- and loading-range in bio-implant wireless power transfer," *IEEE Transactions on Biomedical Circuits and Systems*, vol. 9, no. 6, pp. 875–884, December 2015.
114. C. Kim, J. Park, A. Akinin, S. Ha, R. Kubendran, H. Wang, P.P. Mercier, and G. Cauwenberghs, "A fully integrated 144 MHz wireless-power-receiver-on-chip with an adaptive buck-boost regulating rectifier and low-loss H-Tree signal distribution," *2016 IEEE Symposium on VLSI Circuits* (VLSI-Circuits), Honolulu, HI, pp. 1–2, 2016.
115. L. Cheng, W. Ki, T. Wong, T. Yim, and C. Tsui, "21.7 A 6.78MHz 6W wireless power receiver with a 3-level 1× / ½×/ 0× reconfigurable resonant regulating rectifier," *2016 IEEE International Solid-State Circuits Conference* (ISSCC), San Francisco, CA, pp. 376–377, 2016.
116. L. Cheng, W. Ki, and C. Tsui, "A 6.78-MHz single-stage wireless power receiver using a 3-mode reconfigurable resonant regulating rectifier," *IEEE Journal of Solid-State Circuits*, vol. 52, no. 5, pp. 1412–1423, May 2017.
117. L. Cheng, W. Ki, and C. Tsui, "A wireless power receiver with a 3-level reconfigurable resonant regulating rectifier for mobile-charging applications," *2017 22nd Asia and South Pacific Design Automation Conference* (ASP-DAC), Chiba, pp. 33–34, 2017.
118. C. Kim, S. Ha, J. Park, A. Akinin, P. P. Mercier, and G. Cauwenberghs, "A 144-MHz fully integrated resonant regulating rectifier with hybrid pulse modulation for mm-sized implants," *IEEE Journal of Solid-State Circuits*, vol. 52, no. 11, pp. 3043–3055, November 2017.
119. C.J. Yim and S.H. Park, "Fully integrated direct regulating rectifier with resonance frequency shift for wireless power receivers," *Journal of Semiconductor Technology and Science*, vol. 17, no.5, pp. 597–602, October, 2017.
120. X. Ge, L. Cheng, and W. Ki, "A 13.56 MHz one-stage high-efficiency 0X/1X R3 rectifier for implatable medical devices," *2017 IEEE International Symposium on Circuits and Systems* (ISCAS), Baltimore, MD, pp. 1–4, 2017.
121. A. Hassan, A. Mahar, Naveed, Y. Siddiqi, and S.A. Jawed, "A 200MHz wireless power transfer system using highly efficient regulating reconfigurable rectifier with backscattering communication," *2017 First International Conference on Latest trends in Electrical Engineering and Computing Technologies* (INTELLECT), Karachi, pp. 1–5, 2017.

122. I. Ghotbi, A. Esmailiyan, S.J. Ashtiani, and O. Shoaei, "Extended coupling-range wireless power transfer using 0× / 4× resonant regulating rectifier," *2016 IEEE 59th International Midwest Symposium on Circuits and Systems* (MWSCAS), Abu Dhabi, pp. 1–4, 2016.
123. L. Cheng, X. Ge, W. Ki, and C. Tsui, "A simplified PWM controller for wireless power receiver using a 3-mode reconfigurable resonant regulating rectifier," *2018 IEEE Asia Pacific Conference on Circuits and Systems* (APCCAS), Chengdu, pp. 473–475, 2018.
124. C. Kim, J. Park, S. Ha, A. Akinin, R. Kubendran, P.P. Mercier, and G. Cauwenberghs, "A 3×3 mm^2 fully integrated wireless power receiver and neural interface system-on-chip," *IEEE Transactions on Biomedical Circuits and Systems*, vol. 13, pp. 1736–1746, 2019.
125. [155] B. Lee and D. Ahn, "Robust Self-Regulated Rectifier for Parallel-Resonant Rx Coil in Multiple-Receiver Wireless Power Transmission System," in *IEEE Journal of Emerging and Selected Topics in Power Electronics.* doi: 10.1109/JESTPE.2019.2929279 (Early Access)
126. K. Li, S. Tan, and R.S.Y. Hui, "Single-switch-regulated resonant WPT receiver," *IEEE Transactions on Power Electronics*, vol. 34, no. 11, pp. 10386–10391, November 2019.

7 Rectennas for WPT Systems

7.1 INTRODUCTION

The rectenna design to transfer the power through radio frequency (RF) signals is a very interesting research concept [1], because it can convert RF energy to DC power and plays an important role in wireless power transmission (WPT) [2]. Hence, the rectenna is used in various applications like RF energy harvesting, space energy transfer [3], wireless power transfer [4], satellite application [5], and biomedical telemetry [6]. Nowadays, compact, low-cost, and high-performance rectenna is highly desirable [7]. The rectenna basically consists of receiving antenna, matching network/band-pass filter, rectifying diode, and DC-pass filter. The selection of antenna has a great effect on the system's complexity. Patch antenna with a probe feed can share same ground for rectifying circuit and antenna due to multilayer architecture, thereby reducing the coupling between the antennas and rectifying circuit to some extent and reducing size [7]. For patch antenna fed with a microstrip line, antenna and rectifying circuit will be on the same plane [8]. The input band-pass filter provides impedance matching between the antenna and diodes and rejects unwanted harmonics that are created by the non-linear diodes [9].

7.2 SINGLE-BAND RECTENNAS

Various literatures have discussed about different rectennas operating at different frequencies, such as 2.25GHz [3], X-band [4], 2.45GHz ISM [6], 900MHz [10], 24GHz mm wave band [11], K-band [12], C-band [5], 94GHz mm wave band [13]. Heikkinen et al. [14] have designed three planar rectennas on different PCB materials of FR4 and Duroid RT 6010 and RT5870 at 2.45GHz. The smallest antenna is realized with Duroid RT 6010; however, narrow bandwidth, high manufacturing accuracy requirements, and very high price reduce its general usability. The performance and efficiency of three planar rectennas that are a combination of square patch antenna and a voltage-doubler rectifier circuit, designed on three different PCB materials of RT5870, FR4, and RO3010, have been evaluated at 2.45GHz by Heikkinen and Kivikoski [15]. The best overall performance is achieved with the low-loss high-frequency laminate RT5870. Selvakumaran et al. [1] have discussed about the design procedure of a low-power rectenna for WPT application. Harouni et al. [2] have proposed an efficient microstrip rectenna operating on 2.45GHz ISM band with high harmonic rejection, in which the receiving antenna with proximity-coupled feeding line is implemented in a multilayer substrate, as shown in Figure 7.1.

The optimized circuit at 2.45 GHz for an input power of 10 dBm exhibits a measured efficiency of 74% at 0.3 mW/cm² power density and an output DC voltage of 2.9 V.

In Ref. [3], S-band GaN-based HPA and rectenna have been designed, developed, and evaluated for space energy transfer applications. Kim et al. [4] have presented a printed rectenna and WPT experiment was performed at 9.5 GHz. Within 50 cm, 2×2 array rectenna shows higher efficiency, and outside 60 cm, 4×4 array rectenna shows higher efficiency. Liu et al. [6] have proposed an implantable rectenna, which consists of a planar inverted-F antenna (PIFA) and a rectifier circuit, for far-field wireless power transfer. A parasitic patch over the human body is added to enhance the wireless power link, and the performance of the wireless power link between the implanted antenna and an external antenna is examined. Zhang et al. [7] have presented a compact planar rectenna with high conversion efficiency in the 2.45 GHz ISM band, which is developed by decomposing a planar rectenna topology into two functional parts and then recombining the two parts into a new topology to make the rectenna size reduction. Adami et al. [8] have presented a 2.45-GHz rectenna on textile for RF power transfer and harvesting, in which thin 350-μm polycotton is used as a flexible substrate. In Ref. [16], a 2.45-GHz wideband harmonic rejection rectenna comprising a microstrip-fed circular ring slot antenna (CRSA) and a series-parallel rectifier (SPR) for wireless power transfer is proposed, in which compact microstrip resonant cell is inserted into the CRSA to obtain harmonic suppression over a wide bandwidth (3–8 GHz). Bellal et al. [9] have reported the design, optimization, and experiments of rectenna circuits on transparent substrates containing a linearly polarized circular loop antenna and a coplanar strip-line (CPS) RF-to-dc rectifier, based on shunt SMS7630 Schottky diode, dedicated for wireless power harvesting and operating at different frequency bands: GSM, UMTS, and Wi-Fi. Almohaimeed et al. [10,17] have demonstrated the design of a rectenna to operate over wide dynamic input power range utilizing an adaptive reconfigurable rectifier to overcome the issue of early breakdown voltage in conventional rectifiers and introduced a depletion-mode field-effect transistor to operate as a switch

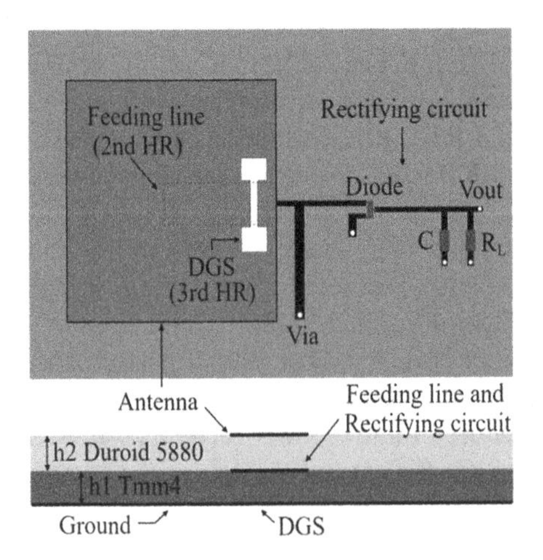

FIGURE 7.1 2.45 GHz rectenna with harmonic rejection. (From Harouni, Z., Osman, L., and Gharsallah, A., *IJCSI Int. J. Comput. Sci. Issues*, 7, 424–427, 2010. With permission.)

and compensate at low and high input power levels for the rectifier. Daskalakis et al. [11] have presented the design of a millimeter-wave rectenna (Figure 7.2) inkjet-printed on glossy photo-paper substrate and integrating an off-the-shelf Schottky diode, and the design has been optimized using harmonic balance optimization combined with electromagnetic simulation in order to maximize the RF-to-DC conversion efficiency.

Takacs et al. have presented K-band [12] and C-band [5] rectenna for satellite applications for powering autonomous wireless sensors. In Ref. [13], a 94-GHz rectenna for high-power application has been developed utilizing a rectifier and 16-element microstrip antenna array combination using fin-line. Lee et al. [18] have presented a compact dual linear polarized cavity-backed patch rectenna with DC power management network for optimized wireless RF power transfer. Badawe and Ramahi [19] have presented a design for a metasurface rectenna that provides near-unity electromagnetic energy harvesting and maximum wireless power transfer to a single load to operate at 2.72 GHz. Almohaimeed et al. [20] have introduced a rectenna by making use of an adaptive rectifier topology and a low-cost metasurface printed antenna which is based on flexible substrates to cover a wide range of input power levels. Khang et al. [21] have proposed the microwave power transfer (MPT) system with the optimal number of rectenna arrays for midrange applications for which a retrodirective power transmitter is designed to overcome the degradation by the near-field effect and enhance the power transfer efficiency. A radiating near-field method of recharging and activating medical implants utilizing a 2.4-GHz rectifying patch antenna (rectenna) is designed by Long et al. [22] using misalignment-insensitive power delivery method.

In Ref. [23], the author investigated the opportunistic beamforming (OBF) scheme in a WPCN where multi-antenna terminals harvest energy from RF signals using a rectenna array and adapt their feedback length based on the harvested power. This type of system can be utilized for future massive network deployments such as machine-to-machine communications and sensor networks, in particular,

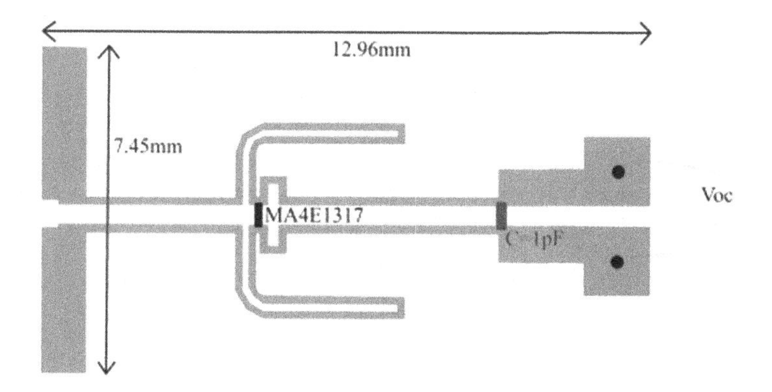

FIGURE 7.2 Fabricated rectenna on commercial photo paper substrate. (From Daskalakis, S., Kimionis, J., Hester, J., Collado, A., Tentzeris, M.M., and Georgiadis, A., "Inkjet Printed 24 GHz rectenna on paper for millimeter wave identification and wireless power transfer applications," *IEEE MTT-S International Microwave Workshop Series on Advanced Materials and Processes* (IMWS-AMP 2017), Pavia, Italy, 20–22 September 2017. With permission.)

the Internet of Things. A large-scale millimeter-wave rectenna and fully integrated rectifiers for far-distance wireless power transfer are reported in Ref. [24].

7.3 BROADBAND AND MULTIBAND RECTENNAS

Wireless energy harvesting (WEH) from ambient electromagnetic sources is becoming emerging technology in which the energy can be directly received from wireless signals, such as the mobile, DTV, and Wi-Fi [25]. So far, the main challenge of the RF energy harvesting is the low power density of ambient RF source. Hence, the DC power obtained is normally very small [26]. With the advent of more and more frequency bands, broadband [27,28] and multiband [29,30] rectennas are desirable to receive or harvest RF power from different sources simultaneously; thus, they outperform the conventional single-band rectennas in terms of overall conversion efficiency as well as total output power.

7.3.1 BROADBAND RECTENNAS

Song et al. [25] have presented a review on recent progress in multiband and broadband rectennas for WEH and wireless power transfer applications. Also, they have developed a hybrid resistance compression technique and a matching network elimination technique to develop rectennas. A novel design concept for a broadband high-efficiency rectenna without using matching networks is presented by Song et al. [26] using an off-center fed dipole antenna with relatively high input impedance over a wide frequency band (Figure 7.3).

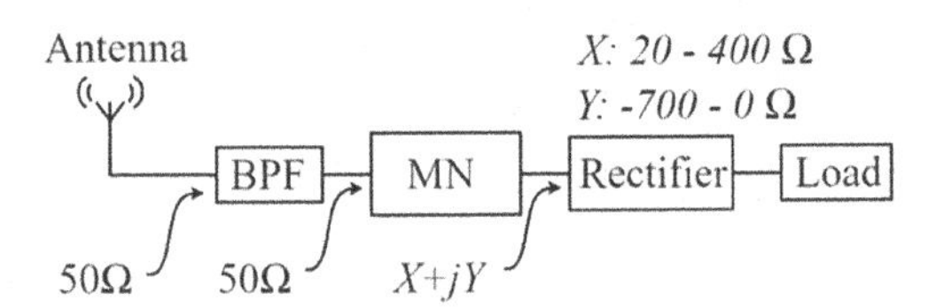

(a) Conventional rectifying antenna system with impedance matching networks

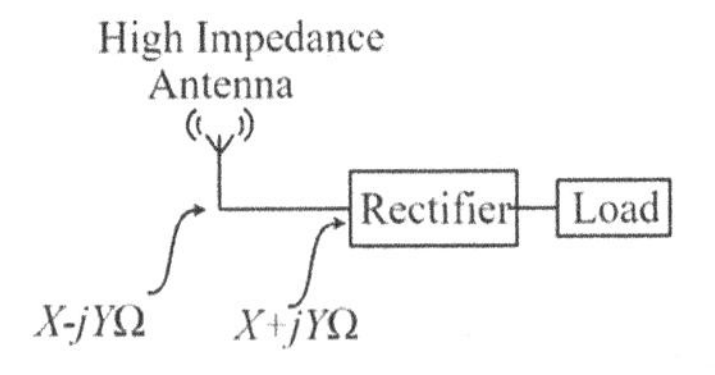

(b) Designed rectifying antenna without using impedance matching networks

FIGURE 7.3 Configuration of rectifying antenna (a) conventional rectifying antenna system with impedance matching networks and (b) designed rectifying antenna without using impedance matching networks. (From Song, C., Huang, Y, Zhou, J., Carter, P., Yuan, S., Xu, Q., and Fei, Z., *IEEE Trans. Indust. Electron.*, 64, 3950–3961, 2017. With permission.)

Chang et al. [27] have designed a wideband slot rectenna, shown in Figure 7.4, comprising a planar wideband slot antenna and a rectifying circuits for WPT operating at the WLAN 2.45 GHz band, fabricated on a cheap FR4 substrate with a compact size of 45 mm by 67 mm. In Ref. [28], a compact and broadband rectifying antenna (rectenna) comprising a novel slotted antenna, broadband dual-stub matching network, and an efficient rectifying circuit is presented for wireless power transfer at LTE-2300/2500 band, in which the ground of the rectifier is directly connected to the ground plane of the antenna for compactness and low energy loss. In Ref. [31], the author presents a novel design method for a rectenna suitable for a wide range of selectable operating frequency band, input power level, and load impedance. The proposed rectenna has a compact size of $90 \times 90 \times 1.58\,\text{mm}^3$ and operates at four different frequency bands that are selectable from 1.1 to 2.7 GHz.

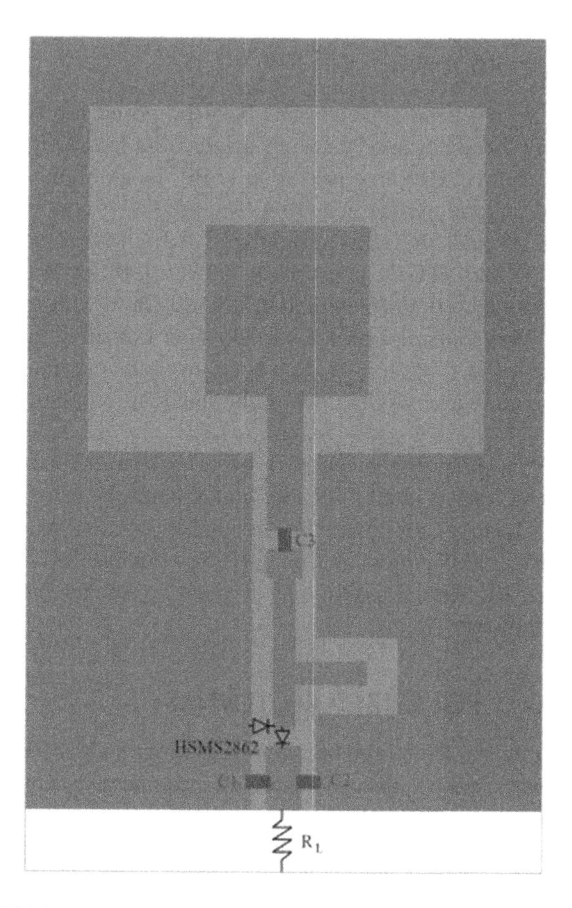

FIGURE 7.4 Wideband rectenna. (From Chang, M., Weng, W., Chen, W., and Li, T., "A wideband planar rectenna for WLAN wireless power transmission," *2017 IEEE Wireless Power Transfer Conference* (WPTC), Taipei, pp. 1–3, 2017. With permission.)

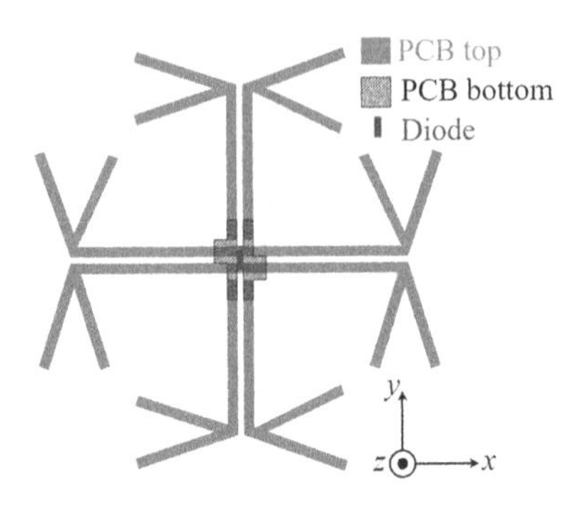

FIGURE 7.5 Ku and K bands rectenna. (From Okba, A., Charlot, S., Calmon, P., Takacs, A., and Aubert, H., "Multiband rectenna for microwave applications," *IEEE Wireless Power Transfer Conference* (WPTC), Aveiro, pp. 1–4, 2016. With permission.)

7.3.2 Multiband Rectennas

In Refs. [29,32], a dual-band rectenna has been proposed for WPT at 2.45- and 5.8-GHz ISM band. The integrated form of a printed dipole antenna, GaAs Schottky barrier diode and a novel CPS low-pass filter (LPF) block higher order harmonics generated from the diode and achieved 84.4% and 82.7% conversion efficiencies in free space at 2.45 and 5.8 GHz, respectively. A high-efficiency dual-band on-chip rectenna at 35 and 94 GHz is proposed in Ref. [30] for WPT. The rectenna comprises a linear tapered slot antenna (LTSA), a finite-width ground coplanar waveguide (FGCPW) transmission lines to slot-line transition, a band-pass filter, and a full-wave rectifier. The measured power conversion efficiencies are 53% and 37% in free space at 35 and 94 GHz, while the incident radiation power density is 30 mW/cm^2.

Okba et al. [33] have reported rectenna (Figure 7.5) that can operate at three operating frequencies 12, 17.6, and 20.2 GHz Ku and K bands by using a multiband four cross dipoles antenna array for powering autonomous wireless sensors. In Ref. [34], an arm-implantable dual-band rectenna comprising a compact PIFA and a rectifier is proposed, in which 402 MHz (MedRadio) is utilized for data telemetry and 915 MHz (ISM) for power transfer.

7.4 CIRCULARLY POLARIZED RECTENNAS

Strict antenna alignment is required between transmitter and receiver to obtain good power conversion efficiency in linearly polarized rectennas [35]. Circularly polarized (CP) rectennas are more suitable for wireless energy transmission to mobile targets because circular polarization avoids changing the output voltage and power conversion efficiency due to the rotation of the transmitter or receiver (Figure 7.6) [36]. Also multiband [37] and broad bandwidth [38] CP rectenna is designed to enhance the output power. In Ref. [35], a CP rectenna has been designed using a truncated corner square patch antenna, a GaAs Schottky diode, input and output low-pass filter for WPT at 5.82 GHz.

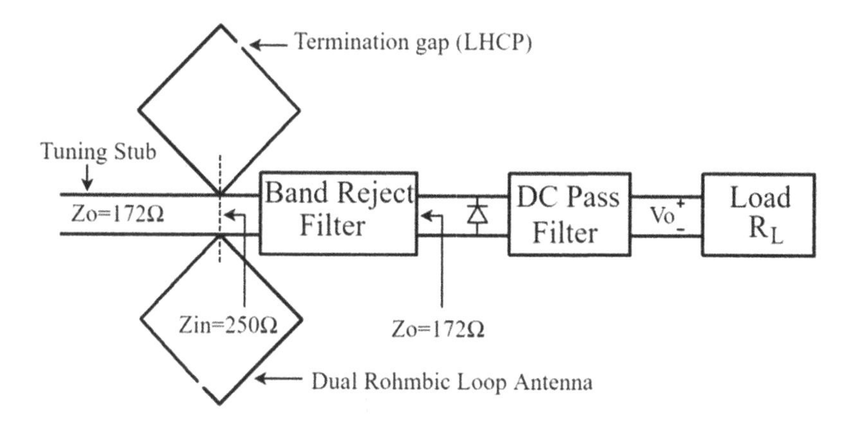

FIGURE 7.6 Rectenna block diagram and spatial orientation. (From Strassner, B. and Chang, K., *IEEE Trans. Microw. Theory Tech.*, 50, 1870–1876, 2002. With permission.)

Strassner et al. [36] have designed a 5.8-GHz CP high-efficiency high-gain rectenna. A (CP) dual-rhombic-loop-antenna (DRLA) with a gain as high as 10.7 dBi and a VSWR of 2 over a 10% bandwidth was used. A CPS band-reject filter (BRF) is employed to suppress the second harmonic generated by the nonlinear diode and prevent its re-radiation. Also, a DC-pass filter (DPF) is used to minimize the leakage of RF power into the resistive DC load. Circular polarization is chosen for the rectenna to maintain constant output DC power irrespective of its orientation and to achieve 82% efficiency at 5.8 GHz.

A dual-band CP rectenna composed of two nested microstrip-fed shorted annular ring-slot antennas and two rectifier circuits for WPT at 2.45 and 5.8 GHz is proposed in Ref. [37]. Output DC voltage of over 2 and 1 V at 2.45 and 5.8 GHz, respectively, was achieved at a transmission distance of 2 m. In Ref. [38], a 5.5. GHz operated CP rectenna is designed in which two rectangular slots diagonally positioned on the square patch antenna to achieve wide axial ratio bandwidth. M/A COM detector diode MA4E1317 is utilized in this design providing a conversion efficiency of 57.3% for a load resistance of 300 Ω. Rectennas operating on 10 GHz with RF-DC conversion efficiency of 75% are studied in Ref. [39], in which a CP quasi-square patches fed by aperture coupling are used as the receiving antennas. The double-layer structure is utilized to minimize the size, decrease the effects of the circuits on the antenna, and improve the bandwidth and gain of the receiving antennas. In Ref. [40],a new CP rectenna whose rectifying circuit includes two diodes is reported. Also, the rectenna consists of a CPS truncated patch antenna and CPS band-pass filter, which can block harmonic signals up to the third order. This dual-diode rectenna achieves an RF-to-DC conversion efficiency of 76% at 5.8 GHz. In Ref. [41], a 2.45-GHz rectifying antenna (rectenna) using a compact dual DCP patch antenna with an RF-DC power conversion part is presented. Dual polarizations are obtained using two crossed slots etched on the ground plane and coupled on a microstrip feed line. Two accesses allow receiving either LHCP or RHCP senses. Due to the coupling feeding technique, no input LPF is needed between the antenna and the rectifier. In Ref. [42], a CP rectenna operating at 35 GHz is reported, in which a pair of side-concave patches is used as a CP receiving antenna with a high gain of 10 dB. An LPF at the input port of the rectifying circuit is used to suppress the high order harmonics, and a DPF is designed to block the fundamental frequency and the harmonics from transmitting to the load. In this rectenna, 81% mW-DC conversion efficiency is achieved. In Ref. [43], the author has discussed about design and implementation of a 24-GHz rectenna for wireless power harvesting and transmission (WPT) techniques in millimeter-wave regime. The proposed structure includes a compact CP substrate integrated waveguide cavity-backed antenna array integrated with a rectifier using Schottky diodes.

In Ref. [44], a rectenna fabricated on a low-cost FR-4 substrate for WPT at 2.45 GHz that contains a DCP patch antenna having high order harmonic rejection property and a pair of rectifying circuits without harmonic rejection filters is proposed. The measured conversion efficiency of the proposed rectenna reaches 82.3% when the inputs of the rectifying circuits are fed at 22 dBm. In Ref. [45], the author has presented an efficient rectenna that contains a new 3×3 antenna array having an operating frequency of 2.45 GHz with circular polarization and high gain, single Zero-Bias Schottky diode rectifier for the high power applications, and an output DC-pass filter associated with the load. An output voltage of 7.02 V and RF-to-DC conversion efficiency of 65.8% can be achieved for an input RF power level of 20 dBm with an optimum load resistor of 0.75 kΩ. In Ref. [46], a rectenna for 2.45 GHz WPT system is proposed which consists of a 3×3 receiving antenna array with circular polarization and high gain of 9.25 dBi, and a microstrip RF-to-DC rectifier based on a voltage doubler circuit with HSMS 2820 Schottky diode. This rectenna provides a high conversion efficiency of 78% and a DC output voltage of 13.6 V for a given input power level of 24 dBm and an optimum load of 950 Ω. In Ref. [47], a dual-band CP rectenna operating at 2.45 and 5.8 GHz is designed for WPT. T-shaped feeding line and a ring slot is utilized for second-order harmonic suppression. In order to improve the rectenna performance, a compact DPF which can smooth the output direct current is used. The simulation results show that the conversion efficiency reaches 75.6% and 71.4% for 2.45 and 5.8 GHz, respectively. In Ref. [48], a compact CP rectenna based on wide-slot broadband antenna is designed at 5.8 GHz to use in mobile applications for WPT. The maximum efficiency of 66.2% and an output DC voltage of 6.33 V have been measured with a 900 Ω load at the power density of 11.4 mW/cm^2. Also 5×5 rectenna array is built, which achieves maximum conversion efficiency of 62%. In Ref. [49], a compact CP rectenna for WPT at 5.8 GHz with harmonic suppression is proposed, in which substrate integrated waveguide structure is adopted for enhanced gain, unidirectional patterns, and surface wave suppression (Figure 7.7). A rectifier, which consists of a matching network, a diode, and a DPF, is integrated with the antenna at the back side having conversion efficiency of 65.87% with a 700 Ω load.

In Ref. [50], a CP rectenna based on slot-coupled microstrip antenna is designed at 35 GHz, in which circular polarization is obtained by cutting the opposite corner of the patch. The axial-ratio (AR) of the presented rectenna is 0.4 dB and the gain is 6.25 dBic at 35 GHz. By designing reasonable MN and output DPF maximum conversion efficiency values of 61% is achieved at an input power of 20 dBm and the load impendence is 139 Ω. In Ref. [51], Huygens LP and CP low-profile and electrically small rectenna systems with high AC-to-DC conversion efficiencies are presented for

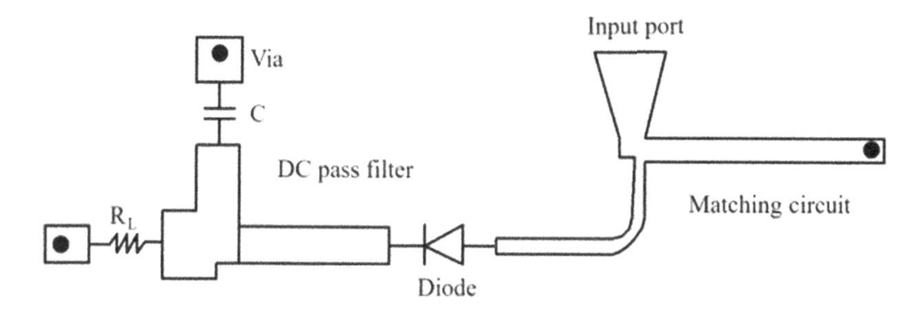

FIGURE 7.7 Schematic and layout of rectifier. (From Yang, Y., Li, L., Li, J., Liu, Y., Zhang, B., Zhu, H., and Huang, K., *IEEE Antennas Wireless Propag. Lett.*, 17, 684–688, 2018. With permission.)

operation in the IMS band around 915 MHz. The HLP system achieved a high 88.9% peak conversion efficiency, and it is suitable for applications in which the polarization of the LP power source and its polarization are easily aligned. The Huygens CP (HCP) peak conversion efficiency was 82%, and it is suitable for applications in which the LP source cannot be aligned to the polarization of the HLP and is ideally suited for WPT applications when the source is also a CP device.

In Refs. [52,53], an electrically small HCP rectenna whose antenna is directly matched to its rectifying circuit is developed in the ISM band at 915 MHz. The electrically small HCP antenna consists of four electrically small near-field resonant parasitic (NFRP) elements: two Egyptian axe dipoles (EADs) and two capacitively loaded loops (CLLs). The rectifier is a full-wave rectifying circuit based on HSMS286C diodes. It is integrated with the HCP antenna on its bottom layer via a CPS without occupying any additional space.

REFERENCES

1. R. Selvakumaran, W. Liu, B. Soong, L. Ming and Y. L. Sum, "Design of low power Rectenna for wireless power transfer," *TENCON 2009–2009 IEEE Region 10 Conference*, Singapore, 2009, pp. 1–5. doi: 10.1109/TENCON.2009.5395942
2. Z. Harouni, L. Osman, and A. Gharsallah, "Efficient 2.45 GHz rectenna design with high harmonic rejection for wireless power transmission," *IJCSI International Journal of Computer Science Issues*, 7, no. 5, pp. 424–427, September 2010.
3. Y. Kobayashi, M. Hori, H. Noji, G. Fukuda, and S. Kawasaki, "The S-band GaN-based high power amplifier and rectenna for space energy transfer applications," *2012 IEEE MTT-S International Microwave Workshop Series on Innovative Wireless Power Transmission: Technologies, Systems, and Applications*, Kyoto, pp. 271–274, 2012.
4. Y. Kim, Y.J. Yoon, J. Shin, and J. So, "X-band printed rectenna design and experiment for wireless power transfer," *IEEE Wireless Power Transfer Conference*, Jeju, pp. 292–295, 2014.
5. A. Takacs, A. Okba, H. Aubert, D. Granena, M. Romier, and A. Bellion, "Compact C-band rectenna for satellite applications," *IEEE Wireless Power Transfer Conference* (WPTC), Montreal, Canada, pp. 1–4, 2018.
6. C. Liu, Y. Guo, H. Sun, and S. Xiao, "Design and safety considerations of an implantable rectenna for far-field wireless power transfer," *IEEE Transactions on Antennas and Propagation*, vol. 62, pp. 5798–5806, 2014.

7. G.F. Zhang, X. Liu, F.Y. Meng, Q. Wu, J.C. Lee, J.F. Xu, C. Wang, and N.Y. Kim, "Design of a compact planar rectenna for wireless power transfer in the ISM band," *International Journal of Antennas and Propagation*, vol. 2014, Article ID 298127, pp. 9, 2014.
8. S. Adami, D. Zhu, Yi Li, E. Mellios, B. H. Stark and S. Beeby, "A 2.45 GHz rectenna screen-printed on polycotton for on-body RF power transfer and harvesting," *2015 IEEE Wireless Power Transfer Conference (WPTC)*, Boulder, CO, 2015, pp. 1–4. doi: 10.1109/WPT.2015.7140161
9. S. Bellal, H. Takhedmit and L. Cirio, "Design and experiments of transparent rectennas for wireless power harvesting," *2016 IEEE Wireless Power Transfer Conference (WPTC)*, Aveiro, 2016, pp. 1–4. doi: 10.1109/WPT.2016.7498848
10. A. M. Almohaimeed, M. C. E. Yagoub and R. E. Amaya, "Efficient rectenna with wide dynamic input power range for 900 MHz wireless power transfer applications," *2016 IEEE Electrical Power and Energy Conference (EPEC)*, Ottawa, ON, 2016, pp. 1–4. doi: 10.1109/EPEC.2016.7771757
11. S. Daskalakis, J. Kimionis, J. Hester, A. Collado, M. M. Tentzeris and A. Georgiadis, "Inkjet printed 24 GHz rectenna on paper for millimeter wave identification and wireless power transfer applications," *2017 IEEE MTT-S International Microwave Workshop Series on Advanced Materials and Processes for RF and THz Applications (IMWS-AMP)*, Pavia, 2017, pp. 1–3. doi: 10.1109/IMWS-AMP.2017.8247367
12. A. Takacs, H. Aubert, L. Despoisse, and S. Fredon, "Design and implementation of a rectenna for satellite application," *2013 IEEE Wireless Power Transfer* (WPT), Perugia, pp. 183–186, 2013.
13. K. Matsui, K. Fujiwara, Y. Okamoto, Y. Mita, H. Yamaoka, H. Koizumi, and K. Komurasaki, "Development of 94GHz microstrip line rectenna," *IEEE Wireless Power Transfer Conference* (WPTC), Montreal, QC, Canada, pp. 1–4, 2018.
14. J. Heikkinen, P. Salonen and M. Kivikoski, "Planar rectennas for 2.45 GHz wireless power transfer," *RAWCON 2000. 2000 IEEE Radio and Wireless Conference (Cat. No.00EX404)*, Denver, CO, USA, 2000, pp. 63–66. doi: 10.1109/RAWCON.2000.881856
15. J. Heikkinen and M. Kivikoski, "Performance and efficiency of planar rectennas for short-range wireless power transfer at 2.45GHz," *Microwave and Optical Technology Letters*, vol. 31, no. 2, pp. 86–91, 2001.
16. Z. Kang, X. Lin, C. Tang, P. Mei, W. Liu, and Y. Fan, "2.45-GHz wideband harmonic rejection rectenna for wireless power transfer," *International Journal of Microwave and Wireless Technologies*, vol. 9, no. 5, pp. 977–983, 2017.
17. A.M. Almohaimeed, M.C.E. Yagoub, and R.E. Amaya, "A highly efficient power harvester with wide dynamic input power range for 900MHz wireless power transfer applications," *2016 16th Mediterranean Microwave Symposium* (MMS), 2016.
18. D.J. Lee, S.T. Khang, S.C. Chae, and J.W. Yu, "Dual linear polarized cavity-backed patch rectenna with DC power management network for optimized wireless RF power transfer," *Microwave and Optical Technology Letters*, vol. 60, pp. 713–717, 2018.
19. M.E. Badawe and O.M. Ramahi, "Efficient metasurface rectenna for electromagnetic wireless power transfer and energy harvesting," *Progress in Electromagnetics Research*, vol. 161, pp. 35–40, 2018.
20. A. M. Almohaimeed, M. C. E. Yagoub, J. A. Lima, R. E. Amaya, G. Xiao and Y. Tao, "Metasurface-Based WPT Rectenna with Extensive Input Power Range in the 900 MHz," *2018 IEEE Canadian Conference on Electrical & Computer Engineering (CCECE)*, Quebec City, QC, 2018, pp. 1–4. doi: 10.1109/CCECE.2018.8447782
21. S.T. Khang, D.J. Lee, I.J. Hwang, T.D. Yeo, and J.W. Yu, "Microwave power transfer with optimal number of rectenna arrays for midrange applications," *IEEE Antennas and Wireless Propagation Letters*, vol. 17, no. 1, pp. 155–159, January 2018.

22. B.J. De Long, A. Kiourti, and J.L. Volakis, "A radiating near-field patch rectenna for wireless power transfer to medical implants at 2.4 GHz," *IEEE Journal of Electromagnetics, RF, and Microwaves in Medicine and Biology*, vol. 2, no. 1, pp. 64–69, March 2018.
23. C. Psomas and I. Krikidis, "A wireless powered feedback protocol for opportunistic beamforming using rectenna arrays," *IEEE Transactions on Green Communications and Networking*, vol. 2, no. 1, pp. 100–113, March 2018.
24. D. Zhao, P. He, and X. Wang, "Millimeter-wave rectenna and rectifying circuits for far-distance wireless power transfer," *2019 12th Global Symposium on Millimeter Waves* (GSMM), Sendai, Japan, pp. 90–92, 2019.
25. C. Song, Y. Huang, J. Zhou and P. Carter, "Recent advances in broadband rectennas for wireless power transfer and ambient RF energy harvesting," *2017 11th European Conference on Antennas and Propagation (EUCAP)*, Paris, 2017, pp. 341–345. doi: 10.23919/EuCAP.2017.7928536
26. C. Song, Y. Huang, J. Zhou, P. Carter, S. Yuan, Q. Xu, and Z. Fei, "Matching network elimination in broadband rectennas for high-efficiency wireless power transfer and energy harvesting," *IEEE Transactions on Industrial Electronics*, vol. 64, no. 5, pp. 3950–3961, May 2017.
27. M. Chang, W. Weng, W. Chen, and T. Li, "A wideband planar rectenna for WLAN wireless power transmission," *IEEE Wireless Power Transfer Conference* (WPTC), Taipei, pp. 1–3, 2017.
28. Y. Shi, Y. Fan, Y. Li, L. Yang, and M. Wang, "An efficient broadband slotted rectenna for wireless power transfer at LTE band," *IEEE Transactions on Antennas and Propagation*, vol. 67, pp. 814–822, 2019.
29. Y.H. Suh and K. Chang, "A novel dual frequency rectenna for high efficiency wireless power transmission at 2.45 and 5.8 GHz," *IEEE MTT-S International Microwave Symposium Digest* (Cat. No.02CH37278), Seattle, WA, vol. 2, pp. 1297–1300, 2002.
30. H. Chiou and I. Chen, "High-efficiency dual-band on-chip rectenna for 35- and 94-GHz wireless power transmission in 0.13-μm CMOS technology," *IEEE Transactions on Microwave Theory and Techniques*, vol. 58, no. 12, pp. 3598–3606, December 2010.
31. C. Song, Y. Huang, P. Carter, J. Zhou, S.D. Joseph, and G. Li, "Novel compact and broadband frequency-selectable rectennas for a wide input-power and load impedance range," *IEEE Transactions on Antennas and Propagation*, vol. 66, no. 7, pp. 3306–3316, July 2018.
32. Y.H. Suh and K. Chang, "A high-efficiency dual-frequency rectenna for 2.45- and 5.8-GHz wireless power transmission," *IEEE Transactions on Microwave Theory and Techniques*, vol. 50, no. 7, pp. 1784–1789, 2002.
33. A. Okba, S. Charlot, P. Calmon, A. Takacs, and H. Aubert, "Multiband rectenna for microwave applications," *IEEE Wireless Power Transfer Conference* (WPTC), Aveiro, pp. 1–4, 2016.
34. S. Bakogianni and S. Koulouridis, "A dual-band implantable rectenna for wireless data and power support at sub-GHz region," *IEEE Transactions on Antennas and Propagation*, vol. 67, no. 11, pp. 6800–6810, November 2019.
35. Y.-H. Suh, C. Wang, and K. Chang, "Circularly polarised truncated-corner square patch microstrip rectenna for wireless power transmission," *Electronics Letters*, vol. 36, pp. 600–602, 2000.
36. B. Strassner and K. Chang, "5.8-GHz circularly polarized rectifying antenna for wireless microwave power transmission," *IEEE Transactions on Microwave Theory and Techniques*, vol. 50, pp. 1870–1876, 2002.
37. J. Heikkinen and M. Kivikoski, "A novel dual-frequency circularly polarized rectenna," *IEEE Antennas and Wireless Propagation Letters*, vol. 2, pp. 330–333, 2003.

38. M. Ali, G. Yang, and R. Dougal, "A new circularly polarized rectenna for wireless power transmission and data communication," *IEEE Antennas and Wireless Propagation Letters*, vol. 4, pp. 205–208, 2005.
39. X. Yang, J. Xu, and D. Xu, "Compact circularly polarized rectennas for microwave power transmission applications," *7th International Symposium on Antennas, Propagation & EM Theory*, Guilin, pp. 1–4, 2006.
40. Y.J. Ren and K. Chang, "5.8-GHz circularly polarized dual-diode rectenna and rectenna array for microwave power transmission," *IEEE Transactions on Microwave Theory and Techniques*, vol. 54, no. 4, pp. 1495–1502, 2006.
41. Z. Harouni, L. Cirio, L. Osman, A. Gharsallah, and O. Picon, "A dual circularly polarized 2.45-GHz rectenna for wireless power transmission," *IEEE Antennas and Wireless Propagation Letters*, vol. 10, pp. 306–309, 2011.
42. Y. Wang and X. Yang, "Design of a high-efficiency circularly polarized rectenna for 35 GHz microwave power transmission system," *Asia-Pacific Power and Energy Engineering Conference*, Shanghai, pp. 1–4, 2012.
43. S. Ladan, A. B. Guntupalli, and K. Wu, "A high-efficiency 24 GHz rectenna development towards millimeter-wave energy harvesting and wireless power transmission," *IEEE Transactions on Circuits and Systems I: Regular Papers*, vol. 61, no. 12, pp. 3358–3366, December 2014.
44. J. Chou, D. Lin, K. Weng, and H. Li, "All polarization receiving rectenna with harmonic rejection property for wireless power transmission," *IEEE Transactions on Antennas and Propagation*, vol. 62, no. 10, pp. 5242–5249, October 2014.
45. M.A. Sennouni, J. Zbitou, B. Abboud, A. Tribak, and M. Latrach, "Efficient rectenna design incorporating new circularly polarized antenna array for wireless power transmission at 2.45GHz," *International Renewable and Sustainable Energy Conference* (IRSEC), Ouarzazate, pp. 577–581, 2014.
46. M.A. Sennouni, J. Zbitou, B. Abboud, A. Tribak, and M. Latrach, "Improved circularly polarized rectenna design for microwave power transmission at 2.45GHz," *International Renewable and Sustainable Energy Conference* (IRSEC), Ouarzazate, pp. 582–586, 2014.
47. Q. Chen, P. Zhao, S. Chen, G. Wang, and X. Chen, "A dual-frequency circularly polarized rectenna for 2.45 and 5.8 GHz wireless power transmission," *IEEE Asia-Pacific Conference on Antennas and Propagation* (APCAP), Auckland, pp. 407–409, 2018.
48. Y. Yang, J. Li, L. Li, Y. Liu, B. Zhang, H. Zhu, and K. Huang, "A 5.8 GHz circularly polarized rectenna with harmonic suppression and rectenna array for wireless power transfer," *IEEE Antennas and Wireless Propagation Letters*, vol. 17, no. 7, pp. 1276–1280, July 2018.
79. Y. Yang, J. Li, L. Li, Y. Liu, B. Zhang, H. Zhu, and K. Huang, "A circularly polarized rectenna array based on substrate integrated waveguide structure with harmonic suppression," *IEEE Antennas and Wireless Propagation Letters*, vol. 17, no. 4, pp. 684–688, April 2018.
50. X. Wang and F. Fan, "Design of a 35 GHz circularly polarized rectenna for wireless power transmission," *International Symposium on Antennas and Propagation* (ISAP), China, pp. 1–3, 2019.
51. W. Lin, R.W. Ziolkowski, and J. Huang, "Electrically small, low-profile, highly efficient, huygens dipole rectennas for wirelessly powering internet-of-things devices," *IEEE Transactions on Antennas and Propagation*, vol. 67, no. 6, pp. 3670–3679, June 2019.
52. W. Lin and R. W. Ziolkowski, "Electrically Small, Highly Efficient, Huygens Circularly Polarized Rectenna for Wireless Power Transfer Applications," *2019 13th European Conference on Antennas and Propagation (EuCAP)*, Krakow, Poland, 2019, pp. 1–3.
53. W. Lin and R.W. Ziolkowski, "Electrically small huygens CP rectenna with a driven loop element maximizes its wireless power transfer efficiency," *IEEE Transactions on Antennas and Propagation*, vol. 68, no. 1, pp. 540–545, January 2020.

8 Matching Networks

8.1 INTRODUCTION

The process of radio frequency energy harvesting (RFEH) starts with the collection of incident energy through antennas, then the collected energy is transferred through the impedance matching circuits, and finally, the energy is converted into DC power by AC-to-DC converters and is harvested [1]. The impedance matching is done between the receiving antenna and rectifier circuits to ensure all received energy is transferred to the rectifier. It is essential to use a matching circuit that matches the receiving antenna impedance to the rectifier circuit to achieve maximum power transfer with enhanced efficiency. Several types of matching circuits are used for RFEH systems. The impedance matching circuit is designed to increase the voltage gain and reduce the transmission loss, which means that the impedance seen by the receiving antenna is equal to the impedance of antenna.

For an efficient RFEH system, an efficient rectifier circuit is important to improve the RF-to-DC conversion efficiency. The overall efficiency of an RFEH system depends on the RF-to-DC conversion efficiency of the rectifier and receiving antenna design. The rectifier circuit consists of an impedance matching network (IMN) and a rectifier. Different types of rectifier circuits have been used to enhance the overall efficiency of RFEH systems. The rectifier contains four parts: a matching network, a RF-to-DC converter, a DC-pass filter, and a load. Matching network is the key element of the rectifier to transfer RF-received power to DC converter. In this chapter, various types of impedance matching circuits used for RFEH systems are reviewed and discussed.

8.2 MATCHING NETWORKS FOR RFEH SYSTEMS

The term 'matching circuit,' also known as impedance matching, is or can be justified as a process of matching one impedance or making one impedance look like another. It is important and necessary to match a load to its source impedance of a driving source. There are a wide variety of parts and components and circuits used for the matching circuit. The applications of the matching circuit or impedance matching are listed below:

a. To get the power signal in an amplifier of a receiver.
b. To transfer the power from one stage to another in a transmitter.

The main reasons for reduction in the energy efficiency of an RFEH system are power leakage during transmission and mismatching between the receiver antenna and rectifier circuits. An IMN enables maximum power transfer between the antenna and the load. The rectifier and remaining circuit are considered as load in the RFEH system. When there is an impedance mismatch, the incident wave gets reflected at the load, which leads to reduction in efficiency. A matching network ensures identical impedance between the source and the load. It can also act as a low pass filter to reject higher-order harmonics generated by the rectifying circuit, which can be re-radiated by the antenna creating loss [1,2]. Hence, a matched filter/impedance network is desired between the rectifier and the antenna. An impedance variation of the rectifier with variation in input power as well as the load leads to degradation in power conversion efficiency (PCE). The preferred feature of matching circuits is that they should be able to match the load impedance with the antenna impedance at different frequencies within wideband, load resistance, and input power. They should also have a small form factor and wideband of operation. The main design challenge is that the antenna impedance changes with load and input power. Tuning circuits can be used to change impedance to a desired value. Bandwidth improvement is provided by the second-order matching circuit than the first-order one [3]. But bandwidth decreases rapidly if the order is increased beyond two matching circuit stages. Lumped elements and distributed microstrip lines are used for implementing matching circuits. Compared to the distributed line-based ones, lumped element-based matching circuits have lower Q values, thus offering wider bandwidths. Due to the parasitic effects associated with the lumped elements, they are not preferred at higher frequencies. The commonly used lumped circuits for matching are T-network, pi-network, shunt inductor, L-network, gamma matching network, band pass filter, etc. The commonly used matching network for low-power regimes is a simple two-component L-type matching network which provides impedance match with a minimal loss [4], and an inductor, L, also preboosts the input signal reaching the rectifier. Due to high-quality factor Q, these matching networks generally have narrow bandwidths. L_{mat} and C_{mat} of the matching network are designed with a source impedance of 50-Ω as as mentioned in equation 8.1 to 8.5. Some matching networks and rectifier circuits are shown in Figures 8.1 and 8.2, which have also been reviewed in detail in Ref. [5].

Source resistance can be calculated as follows:

$$R_S = R_{\text{in}}\left(\frac{1}{1+Q^2}\right) \tag{8.1}$$

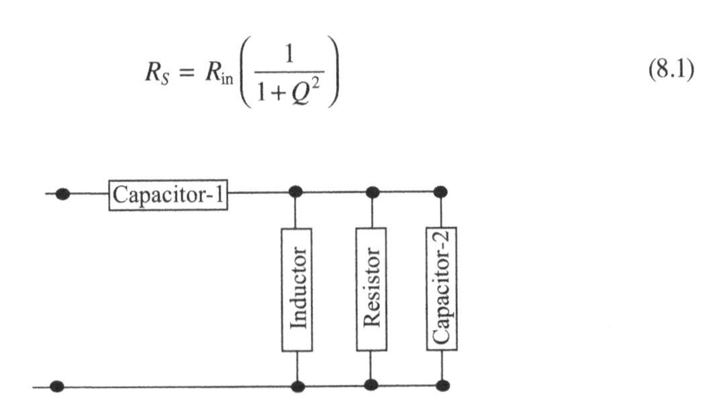

FIGURE 8.1 L-Type matching network. (From Divakaran, S.K., Krishna, D.D., and Nasimuddin, *Int. J. RF Micro. Computer-Aided Eng.*, 29, 1–15, 2019. With permission.)

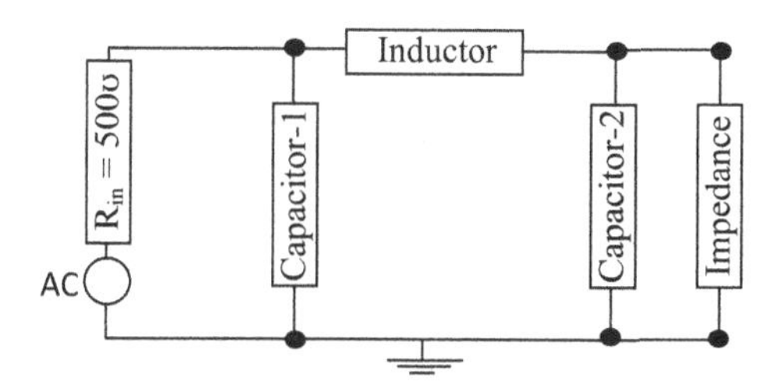

FIGURE 8.2 π-Type matching network. (From Divakaran, S.K., Krishna, D.D., and Nasimuddin, *Int. J. RF Micro. Computer-Aided Eng.*, 29, 1–15, 2019. With permission.)

The quality factor, Q, can be defined as follows:

$$Q = \sqrt{\frac{R_{in}}{R_S} - 1} \tag{8.2}$$

The quality factor expressed in terms of imaginary part of impedance as follows:

$$Q = \frac{Im(Z)}{Re(Z)} \approx \frac{R_{in}}{\omega_0 L_{mat}} - \omega_0 C_{in} R_{in} \tag{8.3}$$

L_{mat} of matching network can be calculated as follows:

$$L_{mat} = \frac{R_{in}}{\omega_0 \left(Q + \omega_0 C_{in} R_{in}\right)} \tag{8.4}$$

C_{mat} is calculated by equating imaginary part to zero as follows:

$$C_{mat} = \frac{R_{in}}{L_{mat}\left(R_{in} - R_S\right)} \frac{1}{\left(\omega_0^2 - \frac{1}{L_{mat} C_{in}}\right)} \tag{8.5}$$

An L-type matching network provides restriction on tuning of two components. This can be overcome by adding an additional L to the circuit transforming it to T- or π-type network. The π-type matching network is shown in Figure 8.2 with a diode replaced by equivalent impedance, and this is superior to the L-type network as it provides an extra degree of freedom with greater amplitude of resonance. The output voltage varies rapidly with frequency compared to the L-type because of the presence of frequency-dependent element.

Design Eq. 8.6 for π-type network is given below:

$$Z_{in} = \left\{\left[\left(R_L - jX_L\right) \| \left(\frac{1}{j\omega C_2}\right)\right] + j\omega L\right\} \| \left(\frac{1}{j\omega C_1}\right) \tag{8.6}$$

The parasitic losses associated with passive elements increase with frequency, and capacitor behavior will change into inductance as frequency increases.

The theoretical limitation of impedance matching bandwidth to parallel load impedance is provided by Bode [6] and Fano [7]:

$$\int_0^{\infty} \ln \frac{1}{|\Gamma(\omega)|} d\omega < \frac{\pi}{R_{\text{load}} C_{\text{load}}} \tag{8.7}$$

For series load,

$$\int_0^{\infty} \frac{1}{\omega^2} \ln \frac{1}{|\Gamma(\omega)|} d\omega < \pi R_{\text{load}} C_{\text{load}} \tag{8.8}$$

where Γ is a reflection coefficient. Γ is ideally 0 for the designed bandwidth range (Δf) and 1 for outside this bandwidth. Hence,

$$\Gamma\omega \ln \frac{1}{|\Gamma(\omega)|_{\min}} < \frac{\pi}{R_{\text{load}} C_{\text{load}}} \tag{8.9}$$

$$\Gamma = e^{-\frac{1}{2\Delta f R_{\text{load}} C_{\text{load}}}} \tag{8.10}$$

We find, according to Bode and Fano, that efficient matching is achieved at the cost of bandwidth. Table 8.1 shows comparison of some matching networks. The T- and π -type matching networks are almost the same.

Broadband impedance matching circuits with two branching sections have been presented in Refs. [8,9]. Radial stub, short stub, and a 6 nH chip inductor are incorporated in the upper branch to obtain impedance matching around the frequency range from 1.8 to 2.5 GHz. The compact rectennas can also be achieved by eliminating matching networks. Song et al. [9] have developed a rectenna system without matching network [9] by changing the antenna to a high impedance antenna that can directly conjugate match with the specific rectifier impedance. The antenna is an off-center-fed dipole antenna operating at the bandwidth of 1.8–2.5 GHz with the imaginary part of impedance varying between 0 and 300 Ω in the desired band. The rectenna was able to achieve a PCE of 75% at 0 dBm input power, which is high compared to other broadband rectennas. An RC low pass filter is incorporated at the output of the rectifier circuits to ensure that pure DC reaches the load. Also, the radial stubs are used to match load variations, and stub-based circuits are being implemented to reduce parasitic losses at high frequencies.

The RFEH circuits are characterized by two metrics: sensitivity and efficiency. The efficiency, expressed as PCE, is the measure of ability of a rectifier to convert

TABLE 8.1
Matching Networks Comparison

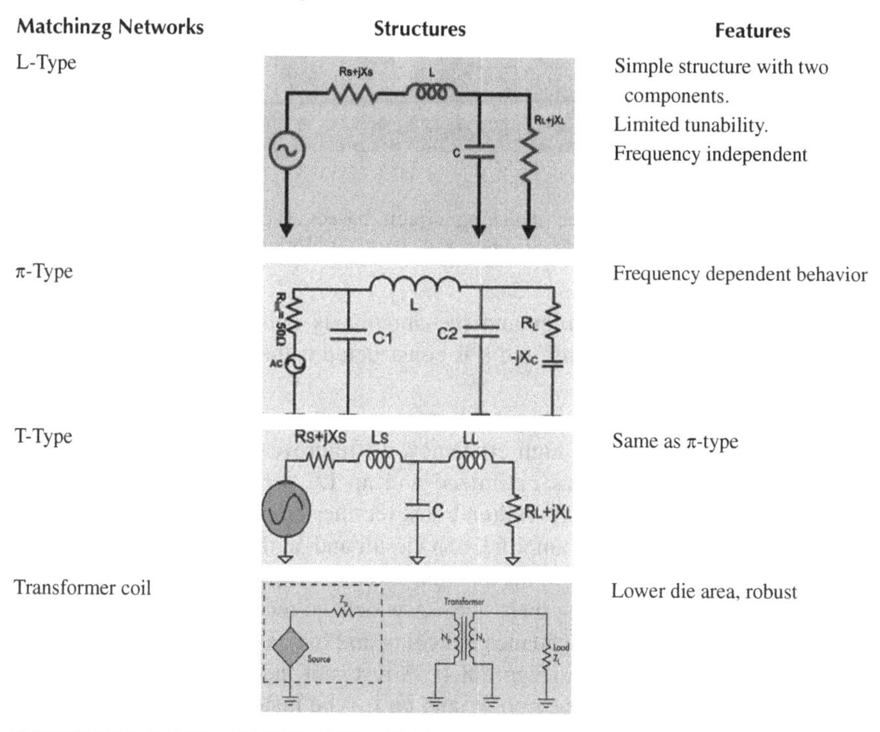

Matchinzg Networks	Structures	Features
L-Type		Simple structure with two components. Limited tunability. Frequency independent
π-Type		Frequency dependent behavior
T-Type		Same as π-type
Transformer coil		Lower die area, robust

the incoming RF energy into DC current. PCE of a diode changes with input power variations. The RF-to-DC conversion efficiency of the rectenna can be calculated as follows:

$$\text{RF-to-DC conversion efficiency} = \frac{\text{output DC power}}{\text{input RF power}} \tag{8.11}$$

The voltage at the input of the rectifier varies according to frequency, which results in variation in the diode impedance leading to efficiency degradation due to mismatch.

An L-slot matching circuit integrated with a high-efficient CP rectenna has been proposed [10] for RFEH system as shown in Figure 8.3. It integrates a simple rectifier circuit with a CP one-sided slot dipole antenna at 2.45 GHz (ISM band) for wireless charging operation at low incident power densities from 1 to 95 μW/cm^2. The rectenna structure is printed on a single-layer, low-cost, commercial FR4 substrate. The integration of the rectifier and antenna produces a low-profile but efficient CP rectenna. To maximize the system efficiency, the matching circuit

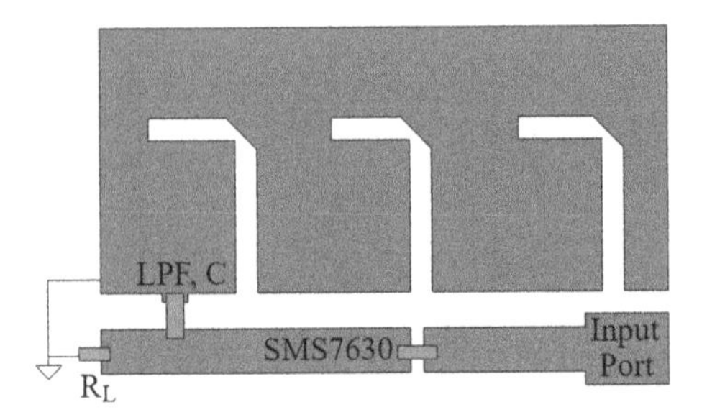

FIGURE 8.3 L-Slot impedance matching circuit based efficient CP rectenna. (From Mansour, M.M. and Kanaya, H., *Electronics*, 8, 1–10, 2019. With permission.)

introduced between the rectifier and the antenna is optimized for a minimum number of discrete components, and it is constructed using multiple L-slot defects in the ground plane.

An RFEH system with an integrated antenna and an impedance matching circuit [11] was investigated for high efficiency. To improve input voltage, the input impedance of the antenna was optimized and an LC series resonance circuit is attached between the Cockcroft–Walton boost rectifier circuit and the antenna. The proposed circuit has a voltage amplification circuit and, at the same time, an impedance matching circuit. Moreover, the antenna and the LC and Cockcroft–Walton circuits are combined by reducing their sizes. A wideband rectifier is combined with a two-level IMN (broadband impedance matching and L-type networks) [12] for making an ambient RFEH system (Figure 8.4). A resonant inductor is used to provide resonance at a specific frequency point, and an L-type matching network was used to tune the input impedance of the circuit without changing the resonance frequency point to facilitate the broadband, while a bandpass matching network was used to achieve impedance matching over the desired broadband.

An RFEH system with an improved dynamic IMN [13] has been proposed to improve efficiency. The system was fabricated on an FR4 substrate using off-the-shelf discrete components, and it is able to convert RF energy to regulated DC voltage in order to supply power to general-purpose electronic devices. The experimental

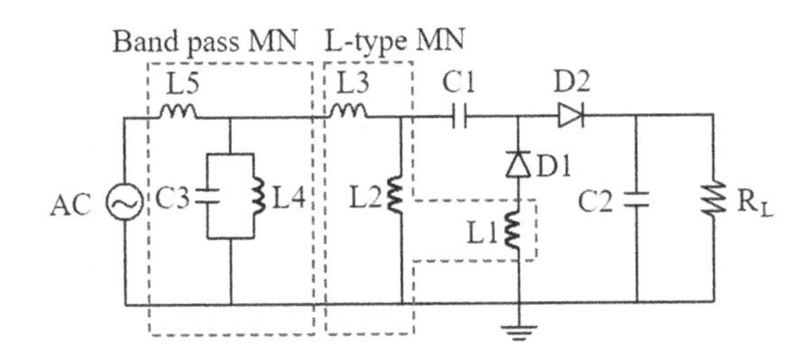

FIGURE 8.4 A two-level IMN-based rectifier. (From Jin, C., Wang, J., Cheng, D.Y., Cui, K.F., and Li, M.Q., *J. Phys. Conf. Series*, 1168, 022020, 2019. With permission.)

results demonstrate the capability of the system to obtain an optimum impedance matching with a received RF power in the range from –10 to +5 dBm. A dual-band bandpass impedance-matching network suitable for a load range from 1 to 10 kΩ has been proposed, which is shown in Figure 8.5 [14].

A broadband two-branch impedance matching, shown in Figure 8.6, has been demonstrated for RFEH system [15]. The upper branch consists of a radial stub, a shorted stub, and a 6 nH chip inductor for impedance matching to 1.8 and 2.5 GHz. The lower branch consists of a bent shorted stub and a 1.8 nH chip inductor for matching to 2.1 GHz. An efficient RFEH system with a dual-band IMN [16] is proposed in Figure 8.7. The circuit comprises a forth-order dual-band impedance matching and a single-series circuit with one double diode, both of which are integrated into a compact shape. The merit of the proposed rectifier circuit is that it can be extended to n number of frequency bands by using only $2 \times n$ matching elements.

A matching technique has been developed in Ref. [17] to increase the efficiency of RFEH systems. The circuit was designed for harvesting energy from ultra-high

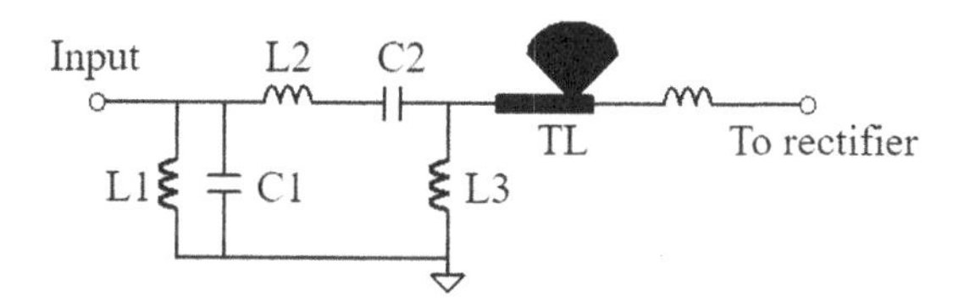

FIGURE 8.5 A dual-band matching network. (From Song, C., Huang, Y., Carter, P., Zhou, J., Yuan, S., Xu, Q., and Kod, M., *IEEE Trans. Antennas Propag.*, 64, 3160–3171, 2016. With permission.)

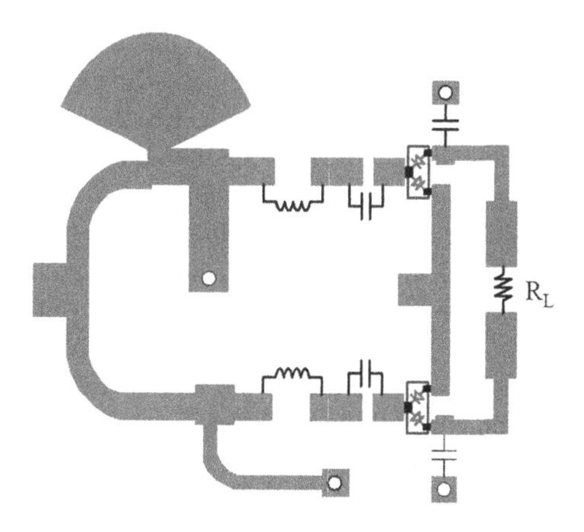

FIGURE 8.6 A rectifier with a two-branch impedance-matching network. (From Song, C., Huang, Y., Zhou, J., Zhang, J., Yuan, S., and Carter, P., *IEEE Trans. Antennas Propag.*, 63, 3486–3495, 2015. With permission.)

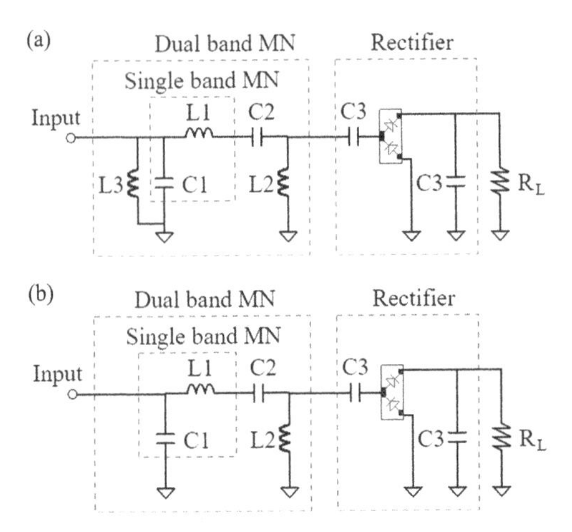

FIGURE 8.7 (a) Circuit diagram of the proposed dual-band rectifier circuit and (b) optimized dual-band rectifier circuit. (From Agrawal, S., Parihar, M.S., and Kondekar, P.N., *Cogent Eng.*, 4, 1332705, 2017. With permission.)

frequency (UHF) with a center frequency of 540 MHz. The multiplier designed in this study successfully covered the channel bandwidth from 535.5 to 541.5 MHz.

8.3 MATCHING NETWORKS FOR WPT SYSTEMS

8.3.1 Matching Networks for Near-Field WPT Systems

Impedance matching techniques are applied in the wireless power transfer (WPT) system for minimizing the losses due to impedance mismatch either at transmitter side or at receiver end (or both), in order to achieve matching at a desired frequency or proper impedance level [18]. For a WPT using magnetic coupled resonant technique, frequencies of the two antennas change according to the air gap in between the antennas. Hence, for a maximum power transfer, the resonance frequencies of the antennas and the frequency of the system should be matched [19]. Therefore, IMNs are needed to adjust the resonance frequency of a pair of antennas at a certain distance. Also, impedance matching from the antenna to load is necessary [20]. Since the efficiency of an inductive power transfer (IPT) system is proportional to the quality factors of the coils, to optimize the geometries of the coupling coils is also a solution to maximize quality factors [21]. Capacitive power transfer (CPT) systems also require matching networks that provide a large voltage or current gain and reactive compensation for high-power large air gap [22]. The conditions for a maximum power transfer from a source antenna to a receiving antenna are examined by Bird et al. [20] when the two antennas are in close proximity, and the results are used to design matching networks between the antenna and a load such as a voltage multiplier for power transfer in a wireless sensor network. The possibility of using IMNs to adjust the resonance frequency of a pair of antennas at a certain distance to 13.56 MHz was studied by Beh et al. [19], where the equivalent circuits are used as reference to calculate the parameters of the impedance matching circuits. Feng et al. [23] have presented a passive matching network that overcomes the threshold-voltage limitation by enhancing the quality factor of the voltage-multiplier front end. The proposed approach significantly increases the powering distance of a 13.56 MHz RF scavenging sensor.

Martins et al. [24] have presented an efficiency analysis and optimization to design multistage matching networks at a single frequency using lumped components by considering complex source and load impedances at each stage of the network. A tri-coil IMN for WPT that provides accurate impedance matching and precise frequency tuning for high coils has been introduced by Ricketts et al. [25] to overcome the challenge of impedance matching the resonators to source/load impedances and adjusting the center frequency of resonance to exactly match the operating frequency. Inagaki [26] has described the theory of image impedance matching for inductively coupled power transfer systems. Lee et al. [27] have presented a new analytic design method for impedance-matched WPT systems using an arbitrary number of coils. The distances between the coils can be varied to maintain the impedance matching performance, which can be determined analytically using the presented design method. In Ref. [18], a network is proposed capable of obtaining matching for different values of coupling at different frequencies.

Kim et al. [28] have presented WPT characteristics according to load variation in multi-device WPT systems using capacitive IMNs and two basis IMNs of series–parallel (SP) and parallel–series (PS) capacitors. Wang et al. [29] have investigated the impact of IMN on transmission efficiency in LED novel driving system based on WPT technology, by using the π-type and L-type matching networks for source and load side, respectively. For a CPT system, impedance matching has been achieved using a Z impedance compensation network with inherent open- and short-circuit immunity [30], a Class-E resonant inverter together with IMN to allow efficient power transfer for 20% variation of the load [31], L-section multistage matching networks that maximize the matching network efficiency for a given number of matching network stages and a given air-gap voltage [22], and L-section multistage matching networks for an optimal number of stages and optimal air-gap voltage [32,33]. In order to improve the distance, power, and efficiency of power transfer in a WPT system under the weak coupling condition, a three-coil weak coupling power transmission model was created by Li et al. [34] using an experiment system with PCB (printed circuit board) coils at a frequency of 6.78 MHz. A comparative study was conducted by Javanbakht et al. [35] to design and analyze the performance of line matching network (LMN) of an IPT system used for charging the batteries of electric vehicles based on fundamental mode analysis (FMA) for the ground assembly, and three LMN configurations were designed based on FMA for the vehicle assembly, namely, CCL, CC, and CLC types. A new method for designing an IMN that realizes the circuit characteristics of an ideal transformer was proposed by Murayama et al. [36] for achieving high efficiency at any point within wide variations of offset and gap and load impedance. Miao et al. [21] have proposed an optimization method of IMNs at the load side to achieve the optimal load, and at the source side to deliver the required amount of power (no-more-no-less) from the power source to the load to maximize the WPT (wireless power transfer) efficiency. Qi et al. [37] have focused on using uniform lines to realize impedance matching and addressed the drawbacks of wireless technology including low efficiency and limited transmission distance using SS (Series-Series) topology resonance network in which two capacitors are included in the wireless network. Kumar et al. [38] have introduced an improved design optimization approach for multistage matching networks comprising L-section stages by allowing these stages to have complex input and load impedances using the method of Lagrange multipliers to determine the gain and impedance characteristics of each stage in the matching network that maximize overall efficiency. Nicoli and Silveira [39] have proposed a general joint design that exploits the advantages of both matching networks and adjustable AC-DC converters by presenting improvements in measured efficiencies of up to 80% (from 9.2% to 16%) at the largest distances, thus maximizing the distance range. Wang et al. [40] have proposed a three-coil wireless power supply system by considering the insulation problem and introduced a method to design the relay coil to be used as an impedance matching window, so that the efficiency and the receiving power can be optimized simultaneously in different application situations. In Ref. [41], a switchless passive dual matching technique was applied to achieve high power transfer efficiency (PTE) over a wide range by generating critical coupling at two other positions.

8.3.2 Matching Networks for Far-Field WPT Systems

For long-distance power transmission, far-field WPT is an interesting option for various practical applications. A microwave/RF rectifier is one of the important components used to convert the transmitted RF energy into DC power in a WPT system. It is generally composed of an input matching network, a nonlinear device, low pass filter, and a DC load [42]. Impedance matching is the most common demand in a WPT system for efficient power conversion. While designing an IMN, it is necessary to know the factors such as operating frequency and available power.

In Ref. [42], the authors present a harmonic-tuned rectifier using a matching network in which an inductor in series with a diode is used to obtain a suitable input reflection coefficient and a suitable harmonic termination. A 2.4 GHz rectifier prototype with an occupied area less than 1.3 cm^2 is implemented to verify the concept, and a measured peak efficiency of 65.4% is obtained at an input power of 0.5 dBm. In Ref. [43], a single stub L-matching network has been utilized in four-stage Villard multiplier to operate in GSM 900 MHz applications. The designed rectifier with a matching circuit enhanced the rectified output power up to 300% during –40 dBm input power compared to the rectifier circuit alone. In Ref. [44], a dual line tapered matching network is designed to match the diode impedance to 50 Ω source impedance at an input power level of –10 dBm and an operating frequency of 2.45 GHz. Even though most of the WPT systems deal with narrowband operation, it is essential to design a rectifier that can accommodate wideband and multiband operations. For this, wideband and multiband matching networks are needed. Broadband matching networks have been designed in Ref. [45] based on the quality factor of input matching circuit having open stub, non-uniform transmission line [46], and multi-stage transmission-line matching [47]. A voltage doubler rectifier having broadband matching was designed by adding two inductors in series with each diode [48]. While a dual-band matching network is designed for voltage doubler rectifier using a four-section matching network [49] and a combination of pie and L-type section [50]. The fluctuations in conditions such as input power, operating frequency, and loading condition can have high effects on the performance of the system; therefore, to overcome these situations, a matching network having extra features is needed. A tunable matching network has been designed using high-quality switched capacitor bank for remote powering of UHF RFIDs and wireless sensor systems, thus extending the load range of the sensor system [51]. In Ref. [52], a varactor-based reconfigurable matching network is designed to overcome the fluctuations in operating conditions such as input power level and frequency. But for a tunable matching network, an extra control unit is needed, which makes the overall circuitry complex. Hence, a matching network is needed that reduces the load and the input power sensitivity without utilizing extra control circuit. Resistance compression networks (RCNs) are a good candidate to overcome this problem. Different matching networks have been designed to compensate the effects of input power variation and load variation on the matching performance. Barton et al. [53] have developed a multi-way transmission-line resistance compression network (TLRCNs) operating at 2.45 GHz, that realized smaller input resistance variations than single-stage designs (Figure 8.8).

The concept of dual-band matching network using RCN [54], impedance compression network (ICN) [55], and TLRCN [56] [57] is introduced, which minimizes the sensitivity of rectifier circuits to variations in input power level and changes in the rectifier load. In Ref. [58], a 915 MHz differential rectifier with a high efficiency of RF-DC conversion over an extended input power range by using an RCN is presented as shown in Figure 8.9. Thus, the loss due to impedance mismatch is reduced, and high efficiency can be achieved in a wider input power range. In Ref. [59], a differential RCN and two ICNs were proposed and applied to the rectifier design. The proposed differential RCN and complex ICNs can be used to reduce the variation ranges of the rectifier's input impedance, which change with the input power level.

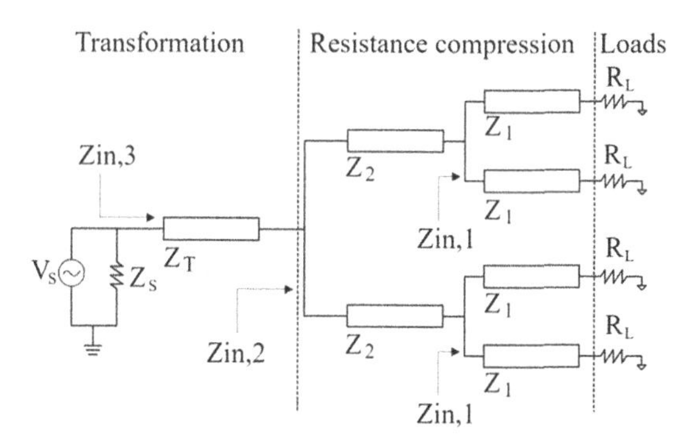

FIGURE 8.8 Four-way TLRCN. (From Barton, T.W., Gordonson, J. and Perreault, D.J., "Transmission line resistance compression networks for microwave rectifiers," *2014 IEEE MTT-S International Microwave Symposium* (IMS2014), Tampa, FL, 1–4, 2014. With permission.)

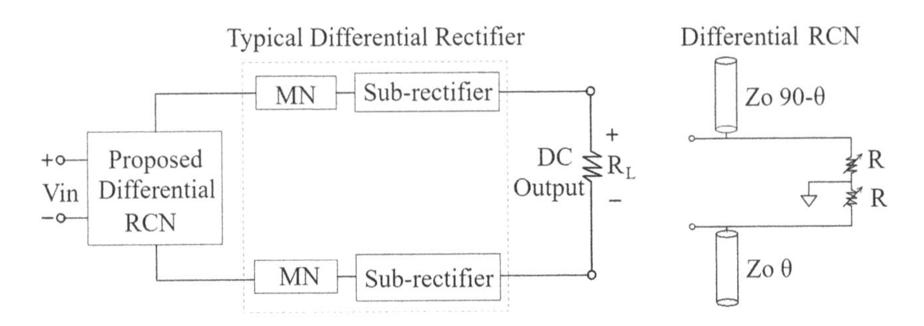

FIGURE 8.9 Differential rectifier with RCN. (From Lin, Q.W. and Zhang, X.Y., *IEEE Trans. Micro. Theory Tech.*, 64, 2943–2954, 2016. With permission.)

The measured input power range for efficiency >50% is from 5.5 to 33.1 dBm for a differential RCN-based rectifier. The input power range for over 50% efficiency of the rectifier with the ICN is 5.2 dB wider than that of the one without the ICN at 2.45 GHz, while it is 3.6 dB wider at 5.8 GHz. Thus, the proposed dual-band ICN can be used to extend the rectifier's input power range at two frequencies. In Ref. [60], a novel rectifier based on a branch-line coupler is proposed to operate within a wide range of input powers, operating frequencies, and output loads.

8.3.3 Adaptive Matching Methods

The efficiency of a WPT system strongly depends on the distance between the transmitter and the receiver [61]. The efficiency drops drastically as the distance between the two antennas increases [62] or if the load changes [63]. For a coupled WPT system, when distance between two coupled resonators changes, it has effects on the coupling coefficient and hence on operating frequency [64]. So, one of the most challenging design issues is to maintain a reasonable level of PTE, even when the load or the distance between the transmitter and the receiver change. To solve this type of problem, an active matching network is necessary to be introduced either at the transmitter side [64] or at both the transmitter and the receiver sides [61].

Adaptive matching methods for a WPT system in the near-field region were investigated by Park et al. [62], and the impedance and resonant frequency characteristic of a near-field power transfer system are analyzed according to coupling distance. A novel algorithm was proposed by Ean et al. [63] for impedance matching and power division considering cross coupling in between two receivers for which the WPT system is first represented by equivalent circuit and an algorithm for impedance matching and power division is derived. A search algorithm based on window prediction was presented by Lim et al. [65] in order to quickly find the best capacitor combination that achieves maximum power transfer, implemented using a resonant frequency of 13.56 MHz, and the experimental results with 1 W power transfer show that the transfer efficiency increases up to 88% when the distance changes from 0 to 1.2 m. A range-adaptive WPT system was proposed by Kim and Jeong [61] to achieve high efficiency over a wide range of distances by using tunable impedance matching techniques and a multi-loop topology to greatly reduce the variation in the input impedance of the WPT system with respect to the distance, where one of the four loops with a different size is selected, depending on the distance. Bito et al. [64] have discussed the feasibility of a real-time active matching circuit for WPT applications, especially for biomedical systems by introducing a genetic algorithm (GA) to optimize the design over a wide range of impedances to match. An efficient free-positioning WPT system for charging multiple devices was proposed by Kim et al. [66] for which a planar transmitting (Tx) coil that consists of a spiral coil, concentric multiple parallel loops, and switchable IMNs at the transmitter was developed. An analytical design of a bias circuit for a real-time active matching circuit utilizing PIN diode switches was discussed in terms of practical circuit restrictions of isolation, response time, and rated DC by Bito and Tentzeris [67] for which an optimized bias circuit under the limitation of use of Arduino Uno micro-controller module was designed. A novel matching circuit design method utilizing a GA and the measured S-parameters of randomly moved coil configurations was discussed by Bito et al. [68] through a detailed comparison of different matching circuit topologies and using a four-cell active matching circuit, which can create 16 different impedance values. Liu et al. [69] have proposed an automated impedance matching system based on L-type IMN and analyzed its effectiveness under various receiver locations using coupled magnetic resonance (CMR). Lu et al. [70] have proposed a load-based design method to improve the efficiency stability of a load-variable system using the efficiency stability coefficient to choose the optimal iterative printed spiral coil (PSCs). Adjustable matching networks in the transmitter were proposed by Yashchenko et al. [71] to compensate the problem of degradation of energy transfer efficiency due to the displacement of the receiver with respect to the transmitter by realizing all components of the considered matching circuit as a set of lumped elements switched by transistors. Wang et al. [72] have used adaptive matching networks for resonant inductive WPT which maximizes the three gains for variable values of the coupling coefficient. Besnoff et al. [73] have introduced a GA approach that avoids the necessity for expensive network measurements or complex calculations by demonstrating the approach in a two-load WPT system that can increase relative efficiency by 23% and 800% to two loads through tuning and does so with a 96.8% reduction in tuning time compared to a brute-force approach. In Ref. [74], to minimize the effect of possible variations of the coupling coefficient and load impedance, a single-side (only one matching network at the input side) and a double-side (two matching networks, one at the input and the other at the output ports of the link) matching schemes are reported. It is described here that only the double-side matching scheme guarantees the achievement of the optimal performance.

8.4.4 Resistance Compression Networks

Soft switching resonant converter provides zero-voltage switching (ZVS) or zero-current switching (ZCS), which can greatly reduce loss at the switching transitions and enabling high efficiency at high frequencies. But many soft-switched resonant designs achieve excellent performance in nominal operating conditions, and their performance can degrade quickly with variations in input and output voltages and power levels [75]. WPT systems are often designed to achieve the highest efficiency at a fixed load value [76] at a fixed coil separation distance and misalignment [77]. A variation in the position of the coils or the load value tends to drastically strike the efficiency, and therefore makes the designed WPT system not practical for applications that are mobile with variable loading conditions such as dynamic wireless charging for electric vehicles or implantable medical device. A new class of matching network known as RCN has been used to minimize the dependency on the variable operating conditions.

Han et al. [78,79] have proposed a new class of matching networks that promise a significant reduction in the load sensitivity of resonant converters and RF amplifiers. The application of resistance compression is demonstrated in a 100 MHz DC-DC converter, which confirms the effectiveness of compression networks in reducing the load sensitivity of resonant DC-DC converters. In Ref. [75], the authors present a new topology for a high-efficiency DC/DC resonant power converter that utilizes an RCN to provide simultaneous ZVS and near-ZCS across a wide range of input voltages, output voltages, and power levels. The RCN maintains desired current waveforms over a wide range of voltages in different operating conditions. The use of ON/OFF control in conjunction with narrowband frequency control enables high efficiency to be maintained across a wide range of power levels. Arteaga et al. [76] presented a current-driven Class D rectifier with an RCN for 6.78 MHz IPT systems, in which the reflected AC load shows a minimal variation for a wide range of DC output loads. The experimental results showed a resistance deviation of 31% for an output DC load varying from 10 to 80 Ω while maintaining high efficiency. The recorded receiving end efficiency ranges from 96.5% to 88.2%. The DC load range was between 10 and 80 Ω, and the RCN was designed to compress this range to an input resistance of 7.5–10.5 Ω. Arteaga et al. [77] presented a novel design approach for loosely coupled IPT systems that can inherently maintain efficient operation against changes in the system's characteristics, coil geometries, and loading conditions. The transmitting end of the proposed IPT system contains a load-independent Class EF inverter that provides a constant amplitude current in the transmitting-end coil and achieves ZVS independent of the coupling factor and the load resistance. A Class D rectifier with an RCN was implemented for the receiving end of the IPT system to ensure that the reflected resistance to the transmitting end is at its optimum value with a minimal dependence on the output load resistance. The combination of the features of the inverter and rectifier allows the IPT system to operate efficiently across a wide range of air gaps, without retuning. Experimental results show a maximum DC-DC efficiency of 83% with a coil separation of one coil diameter and 85 W output power. Choi et al. [80] proposed a high-frequency resonant converter with an ICN to correct horizontal alignment variations between coils in a WPT system. An ICN consists of an RCN and a phase compression network (PCN) to compress both magnitude and phase when a load variation occurs in MRC (magnetic resonance coupling) coils. Ideally, connecting an ICN to MRC coils does not cause additional losses because an ICN consists of only passive components such as inductors or capacitors. As a result, the class Φ2 inverter maintained ZVS and zero dv/dt, and the whole DC-to-DC WPT system implemented with class DE rectifiers provided almost constant efficiency even when misalignments occurred.

REFERENCES

1. L. Marnat, M.H. Ouda, M. Arsalan, K. Salama, and A. Shamim, "RF power harvesting: a review on designing methodologies and application," *Micro and Nano System Letters*, vol. 5, pp. 1–16, 2017.
2. A.R. Lopez, "Review of narrowband impedance-matching limitations," *IEEE Antennas and Propagation Magazine*, vol. 46, no. 4, pp. 88–90, August 2004.
3. T.W. Yoo and K. Chang, "Theoretical and experimental development of 10 and 35 GHz rectennas," *IEEE Transactions on Microwave Theory and Techniques*, vol. 40, pp. 1259–1266, June 1992.
4. Y. Sun and J. Fidler, "Design of impedance matching networks," *IEEE International Symposium on Circuits and Systems*, May 1994.
5. S.K. Divakaran, D.D. Krishna, and Nasimuddin, "RF energy harvesting systems: an overview and design issues," *International Journal of RF and Microwave Computer-Aided Engineering*, vol. 29, pp. 1–15, 2019.
6. W.H. Bode, *Network Analysis and Feedback Amplifier Design*, 1st Edition, Princeton, NJ: Van Nostrand, 1945.
7. R.M. Fano, "Theoretical limitations on the broadband matching of arbitrary impedances," D.Sc. dissertation, Dep. of Elect. Eng., Massachusetts Inst. Technol. (MIT), Cambridge, MA, 1947.
8. C. Song, Y. Huang, J. Zhou, J. Zhang, S. Yuan, and P. Carter, "A high efficiency broadband rectenna for ambient wireless energy harvesting," *IEEE Transactions on Antennas and Propagation*, vol. 63, no. 8, pp. 3486–3495, August 2015.
9. C. Song, Y. Huang, J. Zhou, P. Carter, S. Yuan, Q. Xu, and Z. Fei, "Matching network elimination in broadband rectennas for high efficiency wireless power transfer and energy harvesting," *IEEE Transactions on Industrial Electronics*, vol. 64, no. 6, pp. 3950–3961, May 2017.
10. M.M. Mansour and H. Kanaya, "Novel L-slot matching circuit integrated with circularly polarized rectenna for wireless energy harvesting," *Electronics*, vol. 8, no. 651, pp. 1–10, 2019.
11. Y. Shinki, K. Shibata, M. Mansour, and H. Kanaya, "Impedance matching antenna-integrated high-efficiency energy harvesting circuit," *Sensors*, vol. 7, no. 1763, pp. 1–14, 2017.
12. C. Jin, J. Wang, D.Y. Cheng, K.F. Cui, and M.Q. Li, "A novel wideband rectifier with two-level impedance matching network for ambient wireless energy harvesting," *Journal of Physics: Conference Series*, vol. 1168, pp. 022020, 2019.
13. C. Felini, M. Merenda and F. G. D. Corte, "Dynamic impedance matching network for RF energy harvesting systems," *2014 IEEE RFID Technology and Applications Conference (RFID-TA)*, Tampere, 2014, pp. 86–90. doi: 10.1109/RFID-TA.2014.6934206
14. C. Song, Y. Huang, P. Carter, J. Zhou, S. Yuan, Q. Xu, and M. Kod, "A novel six-band dual CP rectenna using improved impedance matching technique for ambient RF energy harvesting," *IEEE Transactions on Antennas and Propagation*, vol. 64, no. 7, pp. 3160–3171, 2016.
15. C. Song, Y. Huang, J. Zhou, J. Zhang, S. Yuan, and P. Carter, "A high-efficiency broadband rectenna for ambient wireless energy harvesting," *IEEE Transactions on Antennas and Propagation*, vol. 63, pp. 3486–3495, 2015.
16. S. Agrawal, M.S. Parihar, and P.N. Kondekar, "A dual-band RF energy harvesting circuit using 4th order dual-band matching network," *Cogent Engineering*, vol. 4, pp. 1332705, 2017.
17. I. Adam, M.A. Malek, M. Najib, M. Yasin, and H.A. Rahim, "RF energy harvesting with efficient matching technique for low power level application," *ARPN Journal of Engineering and Applied Sciences*, vol. 10, no. 18, pp. 8318–8321, October 2015.

18. M. Dionigi, M. Mongiardo and L. Roselli, "Multi-band design of matched wireless power transfer links," *2014 IEEE Wireless Power Transfer Conference*, Jeju, 2014, pp. 224–227. doi: 10.1109/WPT.2014.6839567
19. Teck Chuan Beh, Takehiro Imura, Masaki Kato and Yoichi Hori, "Basic study of improving efficiency of wireless power transfer via magnetic resonance coupling based on impedance matching," *2010 IEEE International Symposium on Industrial Electronics*, Bari, 2010, pp. 2011–2016.
20. T. S. Bird, N. Rypkema and K. W. Smart, "Antenna impedance matching for maximum power transfer in wireless sensor networks," *SENSORS, 2009 IEEE*, Christchurch, 2009, pp. 916–919. doi: 10.1109/ICSENS.2009.5398165
21. Z. Miao, D. Liuand and C. Gong, "Efficiency enhancement for an inductive wireless power transfer system by optimizing the impedance matching networks," *IEEE Transactions on Biomedical Circuits and Systems*, vol. 11, no. 5, pp. 1160–1170, October 2017.
22. S. Sinha, A. Kumar, S. Pervaiz, B. Regensburger and K. K. Afridi, "Design of efficient matching networks for capacitive wireless power transfer systems," *2016 IEEE 17th Workshop on Control and Modeling for Power Electronics (COMPEL)*, Trondheim, 2016, pp. 1–7. doi: 10.1109/COMPEL.2016.7556756
23. T. Feng and S. Chakrabartty, "Analysis and design of high efficiency inductive power-links using a novel matching strategy," *2012 IEEE 55th International Midwest Symposium on Circuits and Systems* (MWSCAS), Boise, ID, pp. 1172–1175, 2012.
24. G. C. Martins and W. A. Serdijn, "Multistage complex-impedance matching network analysis and optimization," *IEEE Transactions on Circuits and Systems II: Express Briefs*, vol. 63, pp. 833–837, 2016.
25. D.S. Ricketts, M. Chabalko, and A. Hillenius, "Tri-loop impedance and frequency matching with high-resonators in wireless power transfer," *IEEE Antennas and Wireless Propagation Letters*, vol. 13, pp. 341–344, 2014.
26. N. Inagaki, "Theory of image impedance matching for inductively coupled power transfer systems," *IEEE Transactions on Microwave Theory and Techniques*, vol. 62, no. 4, pp. 901–908, April 2014.
27. J. Lee, Y. Lim, H. Ahn, J.D. Yu, and S.O. Lim, "Impedance-matched wireless power transfer systems using an arbitrary number of coils with flexible coil positioning," *IEEE Antennas and Wireless Propagation Letters*, vol. 13, pp. 1207–1210, 2014.
28. J. Kim, D.H. Kim, and Y.J. Park, "Analysis of capacitive impedance matching networks for simultaneous wireless power transfer to multiple devices," *IEEE Transactions on Industrial Electronics*, vol. 62, no. 5, pp. 2807–2813, 2015.
29. T. Wang et al., "Analysis of impedance matching network on LED novel driving system based on the wireless power transfer," *2015 12th China International Forum on Solid State Lighting (SSLCHINA)*, Shenzhen, 2015, pp. 93–96. doi: 10.1109/SSLCHINA.2015.7360697
30. L. Huang, A.P. Hu, A.K. Swain, and Y. Su, "Z-impedance compensation for wireless power transfer based on electric field," *IEEE Transactions on Power Electronics*, vol. 31, no. 11, pp. 7556–7563, November 2016.
31. Y. Yusop, S. Saat, Z. Ghani, H. Husin and S. K. Nguang, "Capacitive power transfer with impedance matching network," *2016 IEEE 12th International Colloquium on Signal Processing & Its Applications (CSPA)*, Malacca City, 2016, pp. 124–129. doi: 10.1109/CSPA.2016.7515817
32. A. Kumar, S. Sinha, A. Sepahvand and K. K. Afridi, "Improved design optimization approach for high efficiency matching networks," *2016 IEEE Energy Conversion Congress and Exposition (ECCE)*, Milwaukee, WI, 2016, pp. 1–7. doi: 10.1109/ECCE.2016.7855189

33. S. Sinha, A. Kumar and K. K. Afridi, "Improved design optimization of efficient matching networks for capacitive wireless power transfer systems," *2018 IEEE Applied Power Electronics Conference and Exposition (APEC)*, San Antonio, TX, 2018, pp. 3167–3173. doi: 10.1109/APEC.2018.8341554
34. Y. Li, K. Song, C. Zhu, G. Wei and R. Lu, "Efficiency optimizing and load matching analysis for the weak-coupling wireless power transfer system using a repeating coil," *2016 IEEE PELS Workshop on Emerging Technologies: Wireless Power Transfer (WoW)*, Knoxville, TN, 2016, pp. 31–34. doi: 10.1109/WoW.2016.7772062
35. P. Javanbakht, G. Liu, M. Abdul-Hak, J. Brunson, O. Cordes and S. Mohagheghi, "Analysis and design of line matching networks for inductive power transfer system of electric vehicles," *2016 IEEE PELS Workshop on Emerging Technologies: Wireless Power Transfer (WoW)*, Knoxville, TN, 2016, pp. 200–207. doi: 10.1109/WoW.2016.7772092
36. T. Murayama, T. Bando, K. Furiya and T. Nakamura, "Method of designing an impedance matching network for wireless power transfer systems," *IECON 2016 - 42nd Annual Conference of the IEEE Industrial Electronics Society*, Florence, 2016, pp. 4504–4509. doi: 10.1109/IECON.2016.7793613
37. C. Qi, H. Xu, Y. Wang, S. Guan and D. Li, "The design of wireless power transmission matching network based on uniform lines," *2017 IEEE International Conference on Systems, Man, and Cybernetics (SMC)*, Banff, AB, 2017, pp. 2556–2560. doi: 10.1109/SMC.2017.8123009
38. A. Kumar, S. Sinha, A. Sepahvand, and K.K. Afridi, "Improved design optimization for high-efficiency matching networks," *IEEE Transactions on Power Electronics*, vol. 33, no. 1, pp. 37–50, 2018.
39. P.P. Nicoli and F. Silveira, "Maximum efficiency tracking in inductive power transmission using both matching networks and adjustable AC-DC converters," *IEEE Transactions on Microwave Theory and Techniques*, vol. 66, no. 7, pp. 3452–3462, July 2018.
40. W. Wang, S. Yang, J. Yang, Q. Wang, and M. Hu, "Optimization analysis of wireless charging system for monitoring sensors overhead the HVPLs based on impedance matching," *IEEE Transactions on Electromagnetic Compatibility*, vol. 61, pp. 1207–1216, 2018.
41. E. Lee, W. Kang, and H. Ku, "A magnetic resonance wireless power transfer system robust to distance variation using dual-matching scheme," *Journal of Electrical Engineering & Technology*, vol. 14, pp. 2097–2103, August 2019.
42. M. Wei, C. Fan, F. Dietrich, and R. Negra, "Compact harmonic-tuned rectifier using inductive matching network," *2019 IEEE MTT-S International Microwave Symposium (IMS)*, Boston, MA, pp. 1515–1518, 2019.
43. I. Adam, M.N.M. Yasin, and M.S. Razalli, "Comparison of rectifier performance using different matching technique," *2016 3rd International Conference on Electronic Design* (ICED), Phuket, pp. 123–127, 2016.
44. A. Eid, J. Costantine, Y. Tawk, M. Abdallah, A.H. Ramadan, and C.G. Christodoulou, "Multi-port RF energy harvester with a tapered matching network," *2017 IEEE International Symposium on Antennas and Propagation & USNC/URSI National Radio Science Meeting*, San Diego, CA, pp. 1611–1612, 2017.
45. H. Sakaki and K. Nishikawa, "Broadband rectifier design based on quality factor of input matching circuit," *2014 Asia-Pacific Microwave Conference*, Sendai, Japan, pp. 1205–1207, 2014.
46. F. Bolos, D. Belo, and A. Georgiadis, "A UHF rectifier with one octave bandwidth based on a non-uniform transmission line," *IEEE MTT-S International Microwave Symposium* (IMS), San Francisco, pp. 1–3, 2016.

47. P. Wu, S.Y. Huang, W. Zhou, W. Yu, Z. Liu, X. Chen, and C. Liu, "Compact high-efficiency broadband rectifier with multi-stage-transmission-line matching," *IEEE Transactions on Circuits and Systems II: Express Briefs*, vol. 66, no. 8, pp. 1316–1320, 2019.
48. H.S. Park and S.K. Hong, "Broadband RF-to-DC rectifier with uncomplicated matching network," *IEEE Microwave and Wireless Components Letters*, vol. 30, no. 1, pp. 43–46, January 2020.
49. M. Aboualalaa, I. Mansour, M. Mansour, A. Bedair, A. Allam, M. Abo-Zahhad, H. Elsadek, K. Yoshitomi, and R.K. Pokharel, "Dual-band rectenna using voltage doubler rectifier and four-section matching network," *2018 IEEE Wireless Power Transfer Conference* (WPTC), Montreal, Canada, pp. 1–4, 2018.
50. J. Liu, X.Y. Zhang, and C. Yang, "Analysis and design of dual-band rectifier using novel matching network," *IEEE Transactions on Circuits and Systems II: Express Briefs*, vol. 65, no. 4, pp. 431–435, 2018.
51. O. Kazanc, F. Maloberti, and C. Dehollain, "High-Q adaptive matching network for remote powering of UHF RFIDs and wireless sensor systems," *IEEE Topical Conference on Wireless Sensors and Sensor Networks* (WiSNet), Austin, TX, pp. 10–12, 2013.
52. T. Ngo and Z. T. Aung, "Varactor-based Rectifier with Adaptive Matching Network for Wireless Power Transfer system," *2018 IEEE Wireless Power Transfer Conference (WPTC)*, Montreal, QC, Canada, 2018, pp. 1–4, doi: 10.1109/WPT.2018.8639391.
53. T.W. Barton, J. Gordonson, and D.J. Perreault, "Transmission line resistance compression networks for microwave rectifiers," *2014 IEEE MTT-S International Microwave Symposium* (IMS2014), Tampa, FL, pp. 1–4, 2014.
54. K. Niotaki, A. Georgiadis, A. Collado, and J.S. Vardakas, "Dual-band resistance compression networks for improved rectifier performance," *IEEE Transactions on Microwave Theory and Techniques*, vol. 62, no. 12, pp. 3512–3521, December 2014.
55. Z. Du and X.Y. Zhang, "High-efficiency single- and dual-band rectifiers using a complex impedance compression network for wireless power transfer," *IEEE Transactions on Industrial Electronics*, vol. 65, no. 6, pp. 5012–5022, June 2018.
56. J. Liu, X.Y. Zhang, and Q. Xue, "Dual-band transmission-line resistance compression network and its application to rectifiers," *IEEE Transactions on Circuits and Systems I*, vol. 66, no. 1, pp. 119–132, January 2019.
57. J. Liu, X.L. Zhao, X.Y. Zhang, and Z.H. Wu, "Microwave rectifier with wide input power range based on resistance/impedance compression networks," *2018 International Applied Computational Electromagnetics Society Symposium - China* (ACES), Beijing, China, pp. 1–2, 2018.
58. Q.W. Lin and X.Y. Zhang, "Differential rectifier using resistance compression network for improving efficiency over extended input power range," *IEEE Transactions on Microwave Theory and Techniques*, vol. 64, no. 9, pp. 2943–2954, September 2016.
59. Y.Y. Xiao, J. Ou, Z. Du, X.Y. Zhang, W. Che, and Q. Xue, "Compact microwave rectifier with wide input power dynamic range based on integrated impedance compression network," *IEEE Access*, vol. 7, pp. 151878–151887, 2019.
60. Zhang, X.Y., Du, Z.X., and Xue, Q., "High-efficiency broadband rectifier with wide ranges of input power and output load based on branch-line coupler," *IEEE Transactions on Circuits and Systems I*, vol. 64, no. 3, pp. 731–739, 2017.
61. J. Kim and J. Jeong, "Range-adaptive wireless power transfer using multi-loop and tunable matching techniques," *IEEE Transactions on Industrial Electronics*, vol. 62, no. 10, pp. 6233–6241, October 2015.
62. J. Park, Y. Tak, Y. Kim, Y. Kim, and S. Nam, "Investigation of adaptive matching methods for near-field wireless power transfer," *IEEE Transactions on Antennas and Propagation*, vol. 59, no. 5, pp. 1769–1773, 2011.

63. K. K. Ean, B. T. Chuan, T. Imura and Y. Hori, "Impedance matching and power division algorithm considering cross coupling for wireless power transfer via magnetic resonance," *Intelec 2012*, Scottsdale, AZ, 2012, pp. 1–5. doi: 10.1109/INTLEC.2012.6374468
64. J. Bito, S. Jeong, and M.M. Tentzeris, "A real-time electrically controlled active matching circuit utilizing genetic algorithms for wireless power transfer to biomedical implants," *IEEE Transactions on Microwave Theory and Techniques*, vol. 64, no. 2, pp. 365–374, 2016.
65. Y. Lim, H. Tang, S. Lim, and J. Park, "An adaptive impedance-matching network based on a novel capacitor matrix for wireless power transfer," *IEEE Transactions on Power Electronics*, vol. 29, no. 8, pp. 4403–4413, 2014.
66. J. Kim, D.H. Kim, and Y.J. Park, "Free-positioning wireless power transfer to multiple devices using a planar transmitting coil and switchable impedance matching networks," *IEEE Transactions on Microwave Theory and Techniques*, vol. 64, no. 11, pp. 3714–3722, 2016.
67. J. Bito and M. M. Tentzeris, "Bias circuit design for a real-time electrically controlled active matching circuit utilizing p-i-n diode switches for wireless power transfer," *2016 IEEE International Symposium on Antennas and Propagation (APSURSI)*, Fajardo, 2016, pp. 405–406, doi: 10.1109/APS.2016.7695911.
68. J. Bito, S. Jeong, and M.M. Tentzeris, "A novel heuristic passive and active matching circuit design method for wireless power transfer to moving objects," *IEEE Transactions on Microwave Theory and Techniques*, vol. 65, no. 4, pp. 1094–1102, April 2017.
69. J. Liu, Y. Zhao, C. Xu and X. Wang, "One-side automated discrete impedance matching scheme for wireless power transmission," *2017 IEEE Wireless Power Transfer Conference (WPTC)*, Taipei, 2017, pp. 1–4. doi: 10.1109/WPT.2017.7953879
70. Y. Lu, S. Mai, C. Zhang, H. Chen and Z. Wang, "Design optimization of printed spiral coils and impedance matching networks for load-variable wireless power transfer systems," *2017 International Conference on Electron Devices and Solid-State Circuits (EDSSC)*, Hsinchu, 2017, pp. 1–2. doi: 10.1109/EDSSC.2017.8126473
71. V. Yashchenko, V. Turgaliev, D. Kozlov, I. Vendik, and A. Katsay, "Adaptive impedance-matching network for wireless power transfer system with off-centre receiver," *Progress in Electromagnetics Research Symposium-Spring* (PIERS), St Petersburg, Russia, 22–25 May, 2017.
72. Q. Wang, W. Che, G. Monti, M. Mongiardo, M. Dionigi, and F. Mastri, "Conjugate image impedance matching for maximizing the gains of a WPT link," *2018 IEEE MTT-S International Wireless Symposium* (IWS), 2018.
73. J. Besnoff, Y. Buchbut, K. Scheim, and D.S. Ricketts, "Dynamic impedance matching of multiple loads in wireless power transfer using a genetic optimization approach," *2018 IEEE/MTT-S International Microwave Symposium*, 2018.
74. Q. Wang, W. Che, M. Dionigi, F. Mastri, M. Mongiardo, and G. Monti, "Gains maximization via impedance matching networks for wireless power transfer," *Progress in Electromagnetics Research*, vol. 164, pp. 135–153, 2019.
75. W. Inam, K.K. Afridi, and D.J. Perreault, "High efficiency resonant DC/DC converter utilizing a resistance compression network," *IEEE Transactions on Power Electronics*, vol. 29, no. 8, pp. 4126–4135, August 2014.
76. J.M. Arteaga, G. Kkelis, D.C. Yates, and P.D. Mitcheson, "A current driven class D rectifier with a resistance compression network for 6.78MHz IPT systems," *2016 IEEE Wireless Power Transfer Conference* (WPTC), Aveiro, pp. 1–4, 2016.
77. J.M. Arteaga, S. Aldhaher, G. Kkelis, D.C. Yates, and P.D. Mitcheson, "Design of a 13.56 MHz IPT system optimised for dynamic wireless charging environments," *IEEE 2nd Annual Southern Power Electronics Conference* (SPEC), Auckland, pp. 1–6, 2016.

78. Y. Han, O. Leitermann, D.A. Jackson, J.M. Rivas, and D.J. Perreault, "Resistance compression networks for resonant power conversion," *IEEE 36th Power Electronics Specialists Conference*, Recife, pp. 1282–1292, 2005.
79. Y. Han, O. Leitermann, D.A. Jackson, J.M. Rivas, and D.J. Perreault, "Resistance compression networks for radio-frequency power conversion," *IEEE Transactions on Power Electronics*, vol. 22, no. 1, pp. 41–53, January 2007.
80. J. Choi, J. Xu, R. Makhoul, and J. Rivas, "Design of a 13.56 MHz dc-to-dc resonant converter using an impedance compression network to mitigate misalignments in a wireless power transfer system," *2018 IEEE 19th Workshop on Control and Modeling for Power Electronics* (COMPEL), Padua, pp. 1–7, 2018.

9 Some Applications

With the advancements in technologies, wireless systems have witnessed a rapid development because of their abilities to simplify the system, and reduce the hardware complexity and system cost. For decades, wireless technology has been abundantly used for communication, and transferring energy to power the other wireless devices. Due to the extensive use of wireless technologies, a significant amount of ambient wireless power is always present around us. The application of wireless technology has expanded to different fields enormously. Thus, the usage of electronic components and the demand for power required to charge them also increased. For effective operation of these devices, continuous supply of power is of prime importance. By the conventional methods, it is not possible to supply power continuously due to the limited lifetime of batteries and inaccessibility of devices in most cases. Due to these limitations of conventional methods, energy harvesting techniques have been considered as a better alternative approach all over the globe. Due to its increased availability in in-door and out-door environments, electromagnetic energy harvesting [1–4] is preferred as one of the alternatives over various forms of energy harvesting techniques such as mechanical [5,6], photovoltaic [7], thermal [8], wind [9], acoustic [10], and hybrid energy harvesting [11–14].

For harvesting electromagnetic energy, a special type of device known as 'rectenna' or rectifying antenna is used. The rectenna is a key component in many wireless applications such as wireless power transmission (WPT), wireless energy harvesting (WEH), and power supply to the implantable devices and wireless sensor nodes. The rectenna was initially designed for WPT applications. The wireless energy transmission concept was first introduced by Brown [15] in the 1960s by combining an antenna and a rectifier in order to receive the high-frequency electromagnetic energy beam. The proposed techniques of the rectenna were demonstrated in 1973 [16] for powering a helicopter through an electromagnetic beam from the earth. The rectenna circuit also finds its implementation in WEH applications. WEH is the process in which the ambient wireless energy is extracted and converted to DC signals to operate low-power electronics devices. For this reason, rectennas are required in energy harvesting systems. However, designing the rectenna for energy harvesting applications poses many challenges. The rectenna is also used in solar power transmission (SPT) applications. It is placed on the earth over a larger area to collect the electromagnetic energy transmitted from the space. The energy collected by the rectenna is converted into DC power, and again this DC power is converted into AC form and then distributed through different transmission lines. The rectenna also finds its applications in wireless sensor networks (WSNs). It is integrated into a sensor element to operate efficiently without any power fluctuation problems. It harvests energy from sources available in the surrounding environment or specially designed sources.

The energy harvesting ability of the rectenna makes it possible to supply power to low-power wireless sensor nodes [17–21], implanted devices [22,23], and the internet of things (IoTs) [24,25]. However, wireless communication occupies the maximum coverage among all existing wireless applications. The other wireless technologies like wireless power transfer (WPT) and WSNs are still in the developing stage. Designing efficient electronic components for these applications is a challenging task that limits their use on a large scale. The performance of the rectenna is greatly affected by some key components such as an antenna and rectifier configuration due to changes in applications and operating frequency. This may also affect the impedance matching of the circuit, resulting in changes in circuit characteristics. The applications of rectenna are likely to be in the range of up to 19.4 THz frequencies [26]; however, at high frequencies, feasible energies are not available in the ambient environment, and hence most of the rectenna applications are confined to low range of microwave frequencies, i.e. up to 10 GHz frequencies. The crucial stage in implementing the rectenna is designing an efficient antenna and a rectifier circuit. The performance of the rectenna usually relies on both the harvesting ability of the antenna and power conversion efficiency (PCE) of the rectifier circuit. In general, the PCE is proportional to the power captured by the antenna. There are various techniques proposed in the literature for increasing the receiving power by an antenna. The common techniques are to use either an array of antennas or multi- or broadband antennas. The antenna array configuration narrows down the bandwidth, thereby reducing the capturing ability and increasing the device performance. This limitation is minimized by a multi- or broadband antenna. The rectifier impedance changes with frequency and amplitude of the harvesting signal, so designing a rectifier circuit with proper matching is quite complicated. As the antenna size reduces, receiving power reduces, so the rectenna must be a sensitive one which responses even to very weak ambient signals as a fraction of μW input power. Thus, picking a suitable diode is most relevant. Therefore, designing a compact rectenna is a challenging task for every researcher.

As far as the literature of review articles regarding rectenna and its applications is concerned, many works have been reported in the existing literature; however, their scope has been very much limited [27–30]. Hamid et al. [27] have presented a review of the energy harvesting system and various rectifier topologies for harvesting purposes. Mrnka et al. [28] have reviewed primarily about various antenna configurations designed for energy harvesting applications. Warda et al. [29] have reviewed the progress in the rectenna, but it is limited to WSNs only. Sleebi et al. [30] have reviewed challenges in the present design of rectenna for energy harvesting applications. Mostly, they have covered various rectenna topologies for either of the rectenna applications only. Thus, a review that simultaneously considers the rectenna design aspects and all rectenna applications such as WPT, WEH, WSN, and implantable devices has been rarely reported to the best of authors' knowledge.

This last chapter is intended to provide comprehensive details on different applications of rectenna. The discussion on applications of the rectenna includes WPT, WEH, WSNs, and implantable devices. A rectenna system usually comprises an antenna, a radio frequency (RF) filter, a matching network, and rectifier circuits.

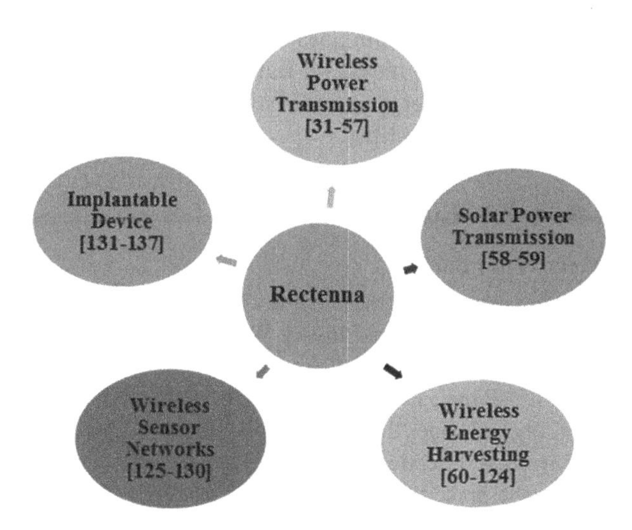

FIGURE 9.1 Major applications of rectennas.

In a rectenna, the antennas are used to sense and receive the RF energy from the surrounding environment, and subsequently, the rectifier circuit converts the received RF signal into the direct current (DC) form. Then, an RF input filter, basically a filter also known as "pre-rectifier," is used to suppress harmonics, and a matching network functions between the filter and the rectifier circuit for a maximum power transfer. The pulsating DC output of the rectifier circuit is smoothened using an output DC pass filter. The major applications of a rectenna system are shown in Figure 9.1.

9.2 RECTENNAS FOR WIRELESS POWER TRANSMISSION

The phenomenon of the transmission of electrical energy from a power source through wireless mode was first conceptualized in 1902. To mitigate the significant losses that occurred in power transmission due to the resistance of electrical wires, Tesla first proposed the WPT technique in 1902 [31]. As mentioned, in the WPT process, electrical energy is transmitted from one place to another without any wire connection. At the transmitting end, the electrical energy is converted into the microwave energy and transmitted to the destination by an antenna. At the receiver end, the microwave energy is received with the help of an antenna and again converted back into the electrical energy. The received energy is converted to DC and stored or delivered to the practical load. The rectenna is used at the receiving end of the WPT system to receive ambient RF signals by an antenna, and the received signal is converted into the DC form by a rectifier circuit. In the past years, several configurations of rectenna systems suitable for WPT systems have been proposed by the researchers. However, rectenna circuits for WPT systems are still under development due to many physical and technical limitations. This section discusses in detail about the advancements that are achieved in rectenna designing techniques over the years. Moreover, it describes how these advancements reduce the power transmission cost and design complexity, and enable the electrical energy storage facility, and it also shows that there is an extreme possibility of power utility.

9.2.1 Single-Band Applications

9.2.1.1 2.45 GHz Wi-Fi Band

Various techniques are reported in the literature to increase transmission efficiency. In 1992, East [32] presented a ground-to-sky WPT system for microwave-powered light aircraft. To provide a stable output from the antenna and enhance the performance, Harouni et al. [33] have fabricated a dual circularly polarized (DCP) antenna with dual-feeding probes. Besides the circular polarization (CP) property, the antenna additionally has harmonic rejection property. However, the use of the dual-probe feeding technique in the presented work increases design complexity. To overcome this limitation, a rectenna with truncated corners and a U-shaped slot embedded on the radiating patch is implemented to obtain CP in Ref. [34]. Moreover, the authors have also implemented the proposed proximity couple feeding technique which is quite complicated. To improve the rectenna performance, a differential feeding for the antenna as well as for the rectifier circuit is investigated in Ref. [35]. The differentially feeding for the antenna improves the antenna gain performance, whereas the differentially driven rectifier exhibits better rectification performance even at low levels of input power. A retrodirective system is implemented using an array of antennas for efficient power transmission in Ref. [36]. These reported designs increase the rectenna dimension, and thus limits rectenna applications. The rectenna with a compact size is desirable. Wang et al. [37] have proposed a technique for WPT using radio frequency energy harvesting (RFEH) in which a folding curve is expanded to seek the antenna size to be compact. Here, it is noticed that a dual-circinal shape and a slotted ground plane improve the performance of antenna and thus the performance of the rectenna system.

9.2.1.2 5.8 GHz Wi-Fi Band/WLAN

In Ref. [38], a coplanar stripline (CPS)-based printed horizontal dipole antenna is designed which exhibits a transmission efficiency of 82% for an input power of 50 mW. A rectenna using a circularly polarized antenna is suggested for improved performance. In Ref. [39], a reflecting surface is placed behind an antenna to achieve more gain. The improvements in gain and CP characteristics, which are made possible by an array of antennas, are discussed in Refs. [40,41]. An array of dual-rhombic-loop antennas are investigated for improving the gain performance of the antenna. A prototype of a circularly polarized antenna is reported in Ref. [40]. In Ref. [42], a CP antenna is implemented with a shorted section introduced into a ring slot embedded with the radiating patch. A rectenna with a wide-slot antenna integrated rectifier circuit at the back of the antenna is fabricated to achieve wide impedance bandwidth with CP characteristics in Ref. [43]. The array of antennas are investigated for further improving the antenna performance. Thereafter, to make the rectenna system alignment independent, retrodirective antennas are desired, in which the receiving antenna automatically aligns towards the power source. A retrodirective antenna with an array of antennas for low-power-density WPT applications is implemented in Ref. [44]. In Ref. [45], an array of two truncated patch antennas are studied for better gain performance in addition to the CP characteristics. A dual-port aperture coupled feeding for the antenna is investigated for designing a rectenna with improved performance in Ref. [46].

9.2.1.3 Other Frequency Band Applications

The polarization factor is one of the important parameters that significantly affects the transmission efficiency. The antenna with improved CP characteristics is desirable to realize a rectenna system, which makes antenna output stable. Thus, in Ref. [47], a rectenna using a circularly polarized antenna is suggested for the improved performance. Mohammod et al. [48] have implemented a CP antenna using two slots positioned diagonally in the microstrip patch radiator. The designed antenna operates at 5.5 GHz frequency. Shabnam et al. [49] have proposed a rectenna with an array of circular patch antennas operating at 24 GHz frequency for achieving high gain in addition to the CP characteristics. The substrate-integrated waveguide (SIW) cavity-backed structure helps in further improving the gain at 3 dB axial ratio bandwidth. In addition to CP characteristics, a wide impedance bandwidth is obtained by choosing a thick substrate for designing the antenna in Ref. [50]. To increase the transmission efficiency with improved rectenna performance, a rectenna with different configurations is investigated at millimeter wave such as single and an array of antenna elements operating at 35 GHz frequency in Ref. [51]. It is observed from the simulated results that the rectenna with an array of series-connected antenna elements provides better DC output voltage than the rectenna with a single antenna element, though the conversion efficiency is less.

9.2.2 Dual-Band Applications

The antenna with multiresonance features can receive more power from the ambient environment than a single-band antenna. The rectenna operating at dual frequencies is proposed for WPT applications for the first time in Refs. [52,53]. In Ref. [52], the antenna is operating at 10 and 35 GHz frequencies. In Ref. [53], the antenna is operating at 2.45 and 35 GHz frequencies. From Refs. [52,53], it is observed that the maximum RF-to-DC conversion efficiency decreases with an increase in operating frequency. Suh et al. [54] have proposed a dual-band rectenna with increased performance using a CPS dipole antenna. A reflecting surface is designed for a better gain performance of the antenna. The dimension of the antenna, as well as the rectenna, increases with an additional reflecting surface. A single-diode rectifier provides better rectification efficiency. A fractal geometry with the Sierpinski triangle technique is implemented for designing an antenna to achieve dual-band characteristics with compact size [55]. An air-gap technique is introduced to improve the gain of the antenna. Hassan et al. [56] have implemented a dual-band antenna with an aperture coupling and air-gap technique for WPT applications.

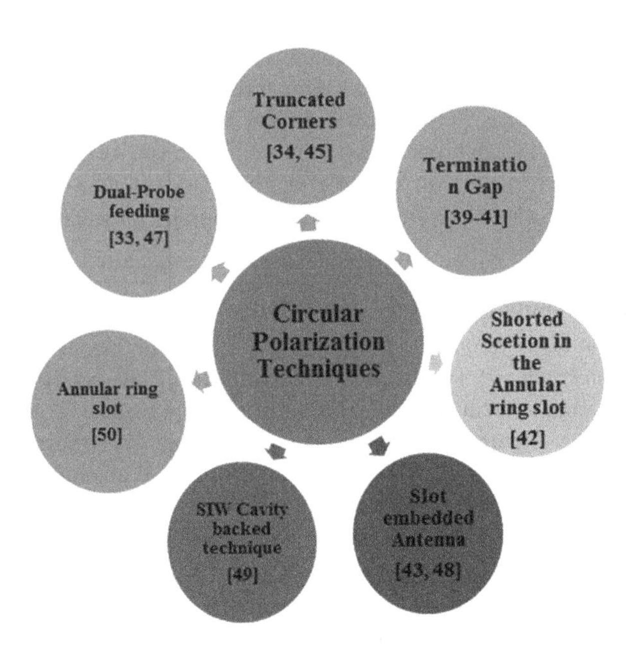

FIGURE 9.2 Various techniques for CP rectennas.

9.2.3 Broadband Antenna Applications

Various techniques have been studied for miniaturization of rectenna for WPT applications. A miniaturized broadband rectenna is achieved with annular and rectangular slots etched on the patch radiator in addition to the slots introduced in the ground plane [57] (Figures 9.2 and 9.3).

9.3 RECTENNAS FOR SOLAR POWER TRANSMISSION

The rectennas have also found their application in SPT systems. In the early 1970s, the rectenna was developed to extract solar energy using multiple energy conversion processes [58]. Initially, a solar satellite was placed in the space to store the solar energy in the form of DC voltage using photovoltaic cells. In the next stage, the DC voltage was converted to microwave radiations and transmitted to the ground station. At the ground station, a very large rectenna array comprising thousands of low-power rectennas were placed to convert the received microwave energy into the DC voltage. An array of dipole antennas operating at 2.45 GHz frequency, low pass filters, and rectifiers constructed using Ga-As Schottky-barrier diodes were used to extract nearly 5 GW energy with a power density of 23 mW/cm^2 at the ground station. The array of each element, the antenna, the filter, and the rectifier were separately constructed and assembled together to form a very large rectenna array occupying a comparatively small effective area. Later, in Ref. [59], the power density of the received power was enhanced by placing a pair of satellites to achieve a conversion efficiency of around 85% with a smaller effective area.

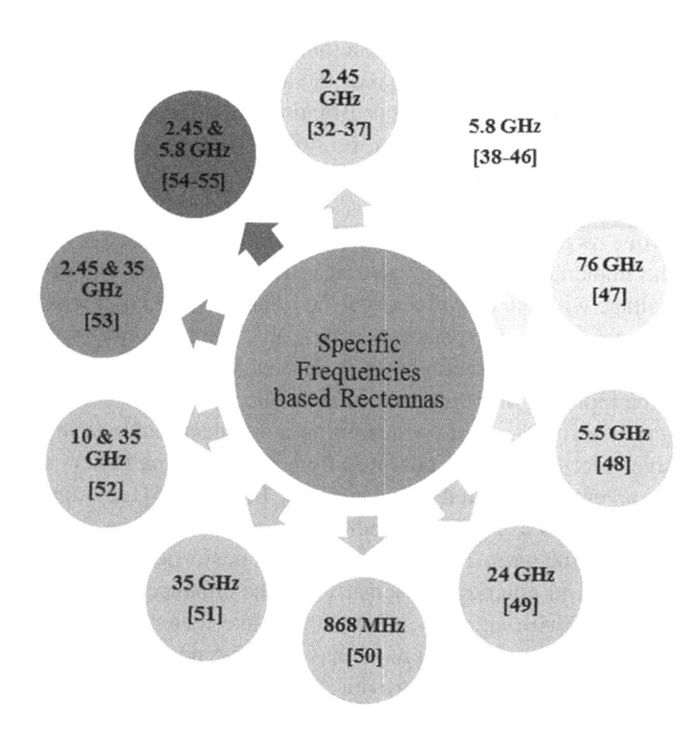

FIGURE 9.3 Rectenna operating bands for WPT applications.

9.4 RECTENNAS FOR WIRELESS ENERGY HARVESTING

The antenna structure is desired to support high gain and achieve CP properties and compactness. Similarly, the rectifier circuit requires to exhibit high PCE in a rectenna system. To meet these requirements of WEH systems, numerous techniques have been proposed to design a rectenna system. In this section, a detailed analysis of these methods is provided based on frequencies of their operations.

9.4.1 Single-Band Applications

9.4.1.1 2.45 GHz Wi-Fi Band Applications

In recent years, the use of Wi-Fi technology and thereby Wi-Fi-enabled devices is increasing abundantly in both indoor and outdoor environments. Thus, the Wi-Fi has become one such technology that emits RF energy continuously. Many rectenna systems are therefore designed to harvest radio energy from the Wi-Fi systems. Paing et al. [60] have implemented a harvesting system using an array of 4×4 spiral antennas operating at 2.4 GHz. However, the array configuration increases system complexity as well as dimensions. Besides, the rectenna system also suffers from polarization sensitivity and does not investigate the gain characteristics. To overcome the design complexity, a rectenna with high gain is proposed for increasing the harvesting power in Ref. [61]. The antenna is operating at 2.45 GHz frequency in the presented design. The authors have used a reflecting surface for improving the antenna gain, which makes the antenna dimension large. Later on, a multilayer antenna design with proper isolation provided between two substrate layers by an air medium is presented in Ref. [62]. The proposed antenna is excited with two probes separated by an annular rectangular ring slot etched on the square radiating plate. Dual-probe feeding increases design complexity. To achieve better performance of a rectenna, a multidirectional radiation technique is investigated in Ref. [63]. The implemented design increases the antenna gain and provides more DC output voltage with the cost of a large size. The proposed methods suffer from the large antenna as well as rectenna dimensions. The rectenna with a compact size is desirable to increase rectenna applications. Various approaches have been proposed in the reported literature to minimize the antenna size. A multi-bending fractal geometry is investigated for designing a compact rectenna in Ref. [64]. A voltage doubler rectifier (VDR) configuration is chosen for better performance. However, the antenna is limited to a low gain. In Ref. [65], asymmetric rectangular conducting strips are used for reducing the antenna size. A tuning stub is used for increasing the gain. The proposed design performance is poor at low RF input power levels and polarization sensitivity. The enhancement in rectenna performance is noticed with an array of reconfigurable rectennas in Ref. [66]. It is observed that a series association of rectennas produces more DC output power, and a parallel association of rectennas provides more conversion efficiency. However, the designed antenna provides a low gain. A rectenna using a Vivaldi antenna and a voltage doubler rectifier configuration is implemented to achieve favorable results such as wideband, compactness, and high conversion efficiency at low input power levels in Ref. [67]. Celik et al. [68] have presented an E-shaped multilayer antenna for WEH applications. Here in the proposed design, compactness is achieved with a combined meander line and defected ground structure techniques. However, the designs presented in Refs. [66–68] suffer from low antenna gain. Techniques that are proposed in the literature suffer either with low antenna gain or polarization sensitivity or lack in performance at low input power levels. Thus, designing a high-gain antenna with a small dimension is preferred for energy harvesting applications. To increase the performance of rectenna by achieving polarization insensitivity, a rectenna is implemented using two crossed slots etched on the ground plane of a microstrip patch antenna in Ref. [69]. The dimension of the antenna is comparatively large. Thus, the rectenna with a compact size, high gain, and insensitive to polarization is desired for energy harvesting applications. A VDR significantly improves the conversion efficiency, as reported in Ref. [70]. In Ref. [71], a rectenna is designed using dual antennas and a full-wave rectifier to maximize the RF-to-DC conversion to 74%. A full-wave bridge rectifier with balanced sources is investigated in Ref. [72]. To achieve favorable rectification results, a Greinacher rectifier topology using an 180^0 hybrid rat-race coupler is investigated in Ref. [73]. The proposed design is quite complicated. A Molybdenum disulfide (MoS2) Schottky-based flexible rectifier circuit is investigated for designing a flexible rectenna system with improved rectenna performance in Ref. [74]. The cut-off frequency of the MoS2 Schottky diode-based rectifier circuit is extended to

10 GHz frequency. Since the cutoff frequency of the diode is one of the important figures of merit, the rectifier is inefficient above the cutoff frequency. Olgun et al. [75] have investigated the DC output performance of a rectenna using two different approaches. The RF power-combining technique is suitable when RF energy harvested by an individual antenna is low, and the combined signal can turn on the diode to be used for conversion. If high power levels received by an individual antenna are applied to an RF power combiner, impedance matching problem may occur, and so DC power-combining technique is suitable when the ambient power level received by an individual antenna is high. In the DC power-combining technique, the series-associated rectenna array yields better performance in terms of the output voltage. However, the conversion efficiency of such series-associated rectenna array is lower than that of parallel-associated rectenna array [76].

9.4.1.2 900/1800 MHz GSM Band Applications

Some other techniques are then proposed for implementing compact rectenna for the desired frequency-based operations. In this subsection, some rectenna designs operating at GSM 900 and 1800 MHz bands are presented. A rectenna with a folded dipole antenna operating at 900 MHz frequency and a voltage doubler rectifier is implemented in Ref. [77]. The antenna becomes compact by folding the two dipole elements. A compact rectenna using square split-ring resonators (SSRRs) and a single-diode coplanar rectifier configuration is implemented in Ref. [78]. The antenna operates over 1.8 GHz frequency band. A modified form of symmetrical Hilbert fractal curve in addition to a slot-loaded ground plane is introduced on either side of the substrate to design the antenna with a tiny size in Ref. [79]. The designed antenna is suitable to harvest energies at 900 MHz frequencies. A Koch fractal geometry with an in-loop-ground plane (ILGP) technique is proposed for designing a compact antenna with better impedance matching in Ref. [80]. The antenna in the present design is operating at 1.8 GHz frequency band. However, the dimension of the antenna is small, and it suffers from low gain. All these antennas reported in the literature are suffered from low gain and polarization sensitivity. A shorted section on a ring-slot antenna perpendicular to the feedline helps in achieving the CP [81]. The proposed design operates at 900 MHz band. The presented antenna suffers from a large antenna dimension.

9.4.1.3 5.8 GHz Wi-Fi/WLAN Band Applications

The enhancement in bandwidth performance of a rectenna is noticed in Ref. [82]. An optimal excitation distribution technique is investigated to increase the impedance bandwidth. Proper adjustment of antennas increases the design complexity. A differential feeding technique is implemented to improve the rectenna performance by increasing the antenna gain in Ref. [83]. A shunt-diode HWR configuration is proposed for better RF-to-DC conversion efficiency. However, in the differential feeding case, two feeding ports are necessary, which increases the design complexity. To overcome this limitation, a slotted patch radiator fed by a coplanar waveguide (CPW) feeding technique is presented in Ref. [84]. A reflecting surface is placed behind an antenna that is used to enhance the gain of the antenna. Thus, the size of the antenna increases.

9.4.1.4 Other Frequency Band Applications

Rectenna circuits have also been designed to operate at different frequencies other than GSM, Wi-Fi, and WLAN bands. An array of planar inverted-F antennas (PIFAs) operating at 2.4 GHz frequency is helpful in increasing the gain is [85]. However, the designed antenna suffers from sensitivity to polarization characteristics. To overcome this limitation, a rectenna with an array of antennas that are embedded with E-shaped slots is investigated to obtain the desired CP characteristics in Ref. [86]. The designed antenna is operating at 2.4 GHz frequency. A 4×4 Buttler matrix is used to implement the overall system, which increases the system complexity. In Ref. [87], four symmetrical strips fabricated along the diagonal directions provide CP characteristics and also increase the impedance bandwidth. The range of impedance bandwidth is 2.53–2.65 GHz. The designed antenna is resonating beyond the Wi-Fi range. At high operating frequencies, the dimension of the antenna is usually small, and it receives less amount of RF input power from the surrounding environment. Thus, harvesting a large amount of power at high frequencies is a challenging task. A rectenna with an array of antennas operating at 35 GHz frequency is implemented in Ref. [88]. The fabricated design increases the gain of the antenna. However, it suffers from a narrow impedance bandwidth. Some literature has also been reported on dielectric resonator antenna (DRA) for energy harvesting purposes. Masius et al. [89] have proposed a rectenna using multilayer DRA energized by a slot-feeding technique for IoT applications.

9.4.2 Dual-Band Applications

The antenna operating in dual bands may harvest a large amount of power when compared to a single-band antenna. The literature based on dual-band antenna that is suitable for energy harvesting applications has been discussed here.

9.4.2.1 1.8 GHz GSM and 2.1 GHz UMTS Bands Applications

In this section, the literature on dual-band rectennas operating at GSM1800 and UMTS2100 bands is explained. A rectenna with an array of quasi-Yagi antennas and HWR configuration is realized to achieve high gain with suitable conversion efficiency in Ref. [90]. The dimension of the antenna is large. In Ref. [91], it is concluded that the rectenna with an array of antenna elements provides better DC output that a single-band CP rectenna. To achieve better gain at dual bands, a rectenna with an array of circular microstrip antennas in the air-cube structure is implemented in Ref. [92]. Designing an antenna is quite complex and increases the dimension also.

9.4.2.2 GSM900 and GSM1800 Bands Applications

The literature on dual-band rectennas operating at GSM900 and GSM1800 bands is explained in this section. An enhancement in rectenna performance is noticed with a rectifier with an extended power range (EPR) technique in Ref. [93]. The proposed EPR technique achieves low threshold voltage with increased breakdown voltage properties. To achieve stable performance from the antenna, a rectenna is designed by introducing an asymmetric step in the ground plane of a monopole antenna operates at dual frequencies in Ref. [94]. However, the antenna suffers from a larger dimension and a low gain. A rectenna with a printed monopole antenna and a VDR configuration is implemented to achieve large conversion efficiency at low input RF power levels in Ref. [95]. The Koch fractal geometry with a Greinacher rectifier topology is implemented to design a compact and efficient rectenna in Ref. [96]. The antennas that are reported in Refs. [95,96] suffer from low antenna gain.

9.4.2.3 Other Dual-Bands Applications

A fractal geometry-based approach is investigated to design a compact rectenna operating at 2.45 and 5.8 GHz frequencies in Ref. [97]. However, the designed antenna is compact, and it suffers from low antenna gain. To achieve better gain, the antenna with a multilayer configuration and a reflecting surface is proposed for the rectenna design in Ref. [98]. The implemented antenna operates at 1.95 and 2.45 GHz frequencies. The dimension of the presented antenna is comparatively large. To mitigate this limitation, a rectenna with aperture-coupled feeding technique is proposed to increase the antenna gain in Ref. [99]. A π-shaped slot introduced on the patch radiator operates at 2.45 and 5.0 GHz frequencies, which holds harmonic rejection property. A rectenna associated with a quasi-Yagi antenna and split ring resonators (SRRs) is presented to achieve a better gain in Ref. [100]. A pair of SRRs integrated into the proposed quasi-Yagi antenna design causes a second resonant frequency. The fabricated design operates at 1.8 and 2.45 GHz frequencies. The rectenna with a small dimension is desirable for energy harvesting applications. The antenna with a Sierpinski triangular fractal-based technique is proposed to design a compact rectenna with effective performance in Ref. [101]. The proposed antenna operates at 2.1 and 5.8 GHz frequencies. The implemented design suffers from polarization sensitivity. A rectenna that is insensitive to polarization is designed by stacking slotted circular patch (SCP) onto the tapered-slit octagon patch (TSOP) in Ref. [102]. The antenna is suitably operated at 0.9 and 2.45 GHz frequencies. The performance of a dual-band rectenna that operates at 2.4 and 5.8 GHz frequencies is investigated in Ref. [103]. The output voltage of rectenna is high with a greater number of voltage multiplier rectifier circuits, especially at high input power levels. A rectenna with a tree-like patch radiator is investigated to operate at 2.45 and 3.5 GHz frequencies in Ref. [104]. A Greinacher rectifier circuit is implemented to obtain increased conversion results at low power levels.

9.4.3 Multiband Applications

A multiband antenna resonates at multiple frequencies. The rectenna with a multiband antenna improves the harvesting ability of the system. Thus, the overall system performance increases. Neeta et al. [105] have proposed a different approach to achieve a stable output from the antenna irrespective of the type of polarization. The authors have implemented the antenna using truncated corners of a radiating patch and cross slots introduced in the patch antenna for CP characteristics at triple bands. A dual-port antenna is implemented to design a high-gain rectenna in Ref. [106]. The designed antenna suffers from a large antenna dimension. To mitigate this limitation, a differential feeding technique is investigated in Ref. [107]. In addition to the differential feeding technique, a reflecting surface is used behind the antenna for further increasing the gain of the antenna. It is noticed that a differential feeding technique provides a larger gain compared to a single feeding technique. However, the impedance bandwidth in the differential feeding technique is comparatively narrower than that in a single feeding case. To increase the performance of a multiband rectenna, a stacked RF harvester technique is proposed in Ref. [108]. The presented design increases the dimension of the system. An efficient multiband rectifier circuit is designed for a compact multiband rectenna system in Ref. [109]. Special types of VDR circuits are designed for each frequency band to get more output from the rectenna in Ref. [110]. A radial stub is used to achieve better impedance matching between the antenna and rectifier circuits. A double-layer metamaterial structure is investigated to produce resonances at four different frequencies that are useful for WEH applications in Ref. [111]. In Refs. [112,113], a six-band rectenna with an improved matching network performance is investigated. In Ref. [112], a frequency-independent log-periodic cross dipole antenna is implemented for six bands with CP characteristics. Whereas in Ref. [113], truncated corners of a radiating patch loaded by several slots are implemented. The presented design is excited by a proximity-coupled technique. In these two rectenna designs, three dual-band impedance matching circuits integrated with rectifiers are distributed in three branches. Each branch is connected to a separate VDR circuit. A six-band rectenna with a fractal geometry is investigated in Ref. [114]. The proposed fractal geometry makes the rectenna size compact.

9.4.4 Broadband Applications

The broadband antennas operate over a wide range of frequencies; thus, the antenna may harvest energy from all available frequencies covered within a wide range of impedance bandwidths. In Ref. [115], it is observed that introducing gradual flaring at each bend of a bent triangular antenna enhances its impedance bandwidth. In Ref. [116], slot introduced in the ground plane of a DRA increases the antenna impedance bandwidth. Besides this, a reflector is used to increase the antenna gain. An array of planar log-periodic dipole array antennas are proposed for the rectenna design in Ref. [117]. The designed antenna provides wide impedance bandwidth in addition to the increase in gain. A notch introduced in a ground plane of a fabricated antenna increases the bandwidth in Ref. [118]. A rectangular patch embedded with a ground plane enhances the bandwidth further. A wideband rectifier is proposed to enhance the performance of a wideband rectenna in Ref. [119]. The proposed rectifier is fabricated on a fractal slotted ground antenna (SGA). To achieve better rectification efficiency, a differential rectifier (DR) topology is implemented in Ref. [120]. A general-purpose diode requires at least a potential drop of 0.3 V for Ge diode and 0.7 V for Si diode to operate and also limited to the low-frequency operation. Thus, it cannot be used at low-power RF. The Schottky [121], high I-V curvature backward tunnel diode [122] and low-barrier Mott [123] diodes are found suitable for WEH systems. A Schottky diode with low turn-on voltage (V_{th}), larger breakdown voltage (V_{br}), smaller series resistance (R_s), and low junction capacitance (C_j) exhibits improved conversion efficiency [124]. However, the backward tunnel diodes are sensitive to the input power level below –30 dBm due to its very low junction capacitance (Figure 9.4).

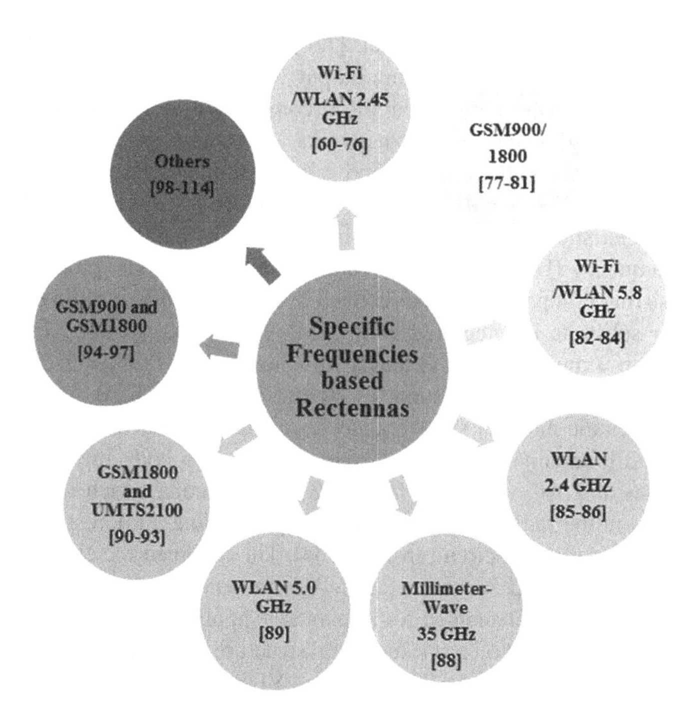

FIGURE 9.4 Rectenna operated frequency bands for RFEH applications.

9.5 WIRELESS SENSOR NETWORKS AND IMPLEMENTABLE DEVICES

Sensors are the primary elements of a WSN which requires an uninterrupted power supply for the low-cost, durable, and continuous functioning of the network. The sensors can be operated by a small amount of power which can be supplied by an in-built rectenna. The use of rectenna to supply power to the sensors remarkably simplifies the architecture of the WSN system. Similarly, a rectenna is also used to supply power to the implanted sensors and monitor their activities in living beings. Since a sensor or an implantable device is of tiny size, it requires a miniature rectenna. In addition, the implantable rectennas require to be biocompatible and need to meet the safety parameters. In this section, different efficiently miniaturized rectennas used for WSN or implantable purposes are discussed.

Tansheng et al. [125] have implemented a rectenna integrated with the Cockcroft–Walton boost converter for sensing and measuring the durability conditions of bridges. It is noticed that the boost converter helps in extending the transmission range with an increased output voltage. As noticed in Ref. [126], introducing an isolation structure between array elements improves the performance of rectennas. Khang et al. [127] have proposed a rectenna for the IoT smart sensor applications by integrating a folded dipole antenna into the voltage multiplier rectifier circuit. A rectenna operating over two Wi-Fi bands is implemented in Ref. [128]. A seven-stage voltage doubler rectifier is realized for increased rectenna output voltage. In Ref. [129], a better PCE is found with a multiband rectifier circuit in comparison to a rectenna circuit comprising multiple single-band rectifier networks. A rectenna circuit consisting of an antenna that resonates at three bands and a single-diode rectifier is investigated in Ref. [130]. A corrugated metal-insulator-metal plasmonic structure is used to obtain tri-band characteristics. Monti et al. [131] have fabricated a rectenna operating in the ultra-high-frequency (UHF) band for wearable applications. The proposed rectenna system consisting of a patch antenna realized by using an adhesive conductive fabric on a bi-layer substrate is integrated with a full-wave bridge rectifier. In Ref. [132], a rectenna with a multilayer PIFA and a full-wave bridge rectifier is implemented for operating a deep-brain stimulation (DBS) device. The proposed planar inverted-F technique makes the antenna size compact. A circularly polarized miniature spiral planar inverted-F antenna (SPIFA) is designed for a deep-body implantable device, which operates at a DC voltage of 0.2V [133]. A rectenna which is insensitive to misalignments in the radiating near field is designed using a microstrip patch and a voltage quadrupler rectifier circuit in Ref. [134]. The designed rectenna is suitable to operate in an insulin pump. A compact triple-band antenna with a voltage doubler rectifier configuration is implemented for implantable applications [135]. The proposed antenna is fabricated based on a stacked spiral structure to make the antenna compact and to obtain desired resonances. The proposed VDR circuit offers better impedance matching and provides better conversion efficiency. Sanusi et al. [136] have fabricated a rectenna to operate dosimeter tags that are mounted on the blood bags. To operate efficiently in the lossy blood environment, an integrated artificial magnetic conductor (AMC) unit cell with a dipole radiator is investigated. A split-ring resonator-based dual-band antenna fabricated on a Kapton polyimide substrate is studied for wearable applications in Ref. [137]. The proposed Kapton is a flexible substrate and physically robust. Most of the rectenna layout area is occupied by the antenna, filter, and the matching circuit. Thus, to limit the size of a rectenna, a miniature but efficient antenna, low-profile filter, and compact matching circuit are required. A traditional antenna is of the half or quarter wavelength in size, which makes it unsuitable for low-frequency WSN and implantable in rectennas operating in low-frequency bands (400MHz to a few GHz ranges). Numerous antennas, filters, and matching circuit miniaturization techniques are reported in the literature. Among those, curly shaped designs [140–145] are most suitable for the WSN and implantable rectennas. In a curly shaped patch technique, the half or quarter wavelength is covered using a meander line [138], spiral [139], etc., and in a stacked form, the length is covered using multilayers connected with vias named as PIFA structure (Tables 9.1–9.3).

9.6 ENVIRONMENT FOR RECTENNA TESTING

Based on the application of the rectenna, the performance can be evaluated either by on-field ambient wireless source or by directional source in the laboratory. While for on-field performance measurement, indoor and outdoor locations are considered. In the on-field measurement, the rectenna is placed in the open space (indoor or outdoor)

TABLE 9.1
Antenna Miniaturization Techniques Comparison

Ref.	Technique	Patch Shape	Substrate Shape	Volume (mm³)
[141]	Curly patch	Spiral	Rectangular	10240
[142]	Curly patch with PIFA	Spiral	Rectangular	1200
[143]	Curly patch with PIFA	Meandered	Rectangular	1200
[144]	Staked curly patch with PIFA	Spiral	Square	273.6
[145]	Staked curly patch with PIFA	Spiral	Square	190

TABLE 9.2
Rectenna Gain Improvement Techniques Comparison

	Max. Gain (dBi)			Fractional Bandwidth (in %)		
Technique	**<4.0**	**≥4.0 & ≤6.0**	**>6.0**	**<10%**	**≥10% & <20%**	**≥20%**
Reflecting surface		[54,107,116]	[39,61,84,98, 107,116]	[61,98,107]	[39]	[54,84,116]
Dual-probe feeding			[46,62,106]	[46,62]	[106]	[106]
Differential feeding		[35,107]	[83,107]	[35,83,107]		
Antenna array		[85,92]	[40,41,45,88, 90,92,117]	[45,88,92]	[85]	[90,117]
SIW cavity-backed			[49]		[49]	
Air-gap			[56,89,92]	[56,92]	[89]	
Aperture coupling			[46,56,99]	[46,56,99]		[99]
SRR		[100]	[100]	[100]	[100]	

TABLE 9.3
Comparison of Various Miniaturization Rectenna Techniques

	Maximum Gain (dBi)			Fractional Bandwidth (in %)		
Technique	**<4.0**	**≥4.0 & ≤6.0**	**>6.0**	**<10**	**≥10 & <20**	**≥20**
Expanding a folding curve	[37,104]	-	-	[37,104]	-	-
Fractal geometry-based	[55,64,79, 80,96,97]	[55]	-	[55,64,79, 80,97]	[96]	[96]
Slotted patch	[39]	-	-	-	-	[39]
Asymmetrical rectangular strips	-	[65]	-	-	-	[65]
Tapered geometry	[67]	-	-	-	-	[67]
Meandered line	[68]	-	-	-	[68]	-
Folded dipole	[77]	-	-	[77]	-	-
SSRR structure	[78]	-	-	[78]	-	-
Planar inverted-F	[132,133]	-	-	[132,133]	-	-

and the DC level is measured using the voltmeter. The on-field measurement is essential for applications such as WEH and WSN. In Refs. [90,112,115], the on-field indoor and outdoor performances are measured. The performance evaluation using a directional source is performed in an anechoic chamber. Usually, a horn antenna is focused on the rectenna system to provide the input power. The performance evaluation using a directional source is discussed. In the reported works, only limited rectenna configurations have been tested in the outdoor atmosphere practically.

REFERENCES

1. S. Agrawal, M.S. Parihar, and P.N. Kondekar, "A dual-band rectenna using broadband DRA loaded with slot," *International Journal of Microwave and Wireless Technologies*, vol. 10, no. 1, pp. 59–66, February 2018.
2. G. Monti, F. Congedo, D.D. Donno, and L. Tarricone, "Monopole-based rectenna for microwave energy harvesting of UHF RFID systems," *Progress in Electromagnetics Research C*, vol. 31, pp. 109–121, 2012.
3. V. Palazzi, J. Hester, J. Bito, F. Alimenti, C. Kalialakis, A. Collado, P. Mezzanotte, A. Georgiadis, L. Roselli, and M.M. Tentzeris, "A novel ultra-lightweight multiband rectenna on paper for RF energy harvesting in the next generation LTE bands," *IEEE Transactions on Microwave Theory and Techniques*, vol. 66, no. 1, pp. 366–379, 2017.
4. A. Khemar, A. Kacha, H. Takhedmit, and G. Abib, "Design and experiments of a dual-band rectenna for ambient RF energy harvesting in urban environments," *IET Microwaves, Antennas & Propagation*, vol. 12, no. 1, pp. 49–55, 2018.
5. H. Okamoto, T. Onuki, S. Nagasawa, and H. Kuwano, "Efficient energy harvesting from irregular mechanical vibrations by active motion control," *Journal of Microelectromechanical Systems*, vol. 18, no. 6, pp. 1420–1431, 2009.
6. X. Li, M. Guo, and S. Dong, "A flex-compressive-mode piezoelectric transducer for mechanical vibration/strain energy harvesting," *IEEE Transactions on Ultrasonics, Ferroelectrics and Frequency Control*, vol. 58, no. 4, pp. 698–703, 2011.
7. P.K.C. Wong, A. Kalam, and R. Barr, "Modeling and analysis of practical options to improve the hosting capacity of low voltage networks for embedded photo-voltaic generation," *IET Renewable Power Generation*, vol. 11, pp. 625–32, 2017.
8. M. Ashraf and N. Masoumi, "A thermal energy harvesting power supply with an internal startup circuit for pacemakers," *IEEE Transactions on Very Large-Scale Integration (VLSI) Systems*, vol. 24, no. 1, pp. 26–37, 2016.
9. Y.K. Tan and S.K. Panda, "Self-autonomous wireless sensor nodes with wind energy harvesting for remote sensing of wind-driven wildfire spread," *IEEE Transactions on Instrumentation and Measurement*, vol. 60, pp. 1367–1377, 2011.
10. C. Mc Caffrey, T. Sillanpaa, H. Huovila, J. Nikunen, S. Hakulinen, and P. Pursula, "Energy autonomous wireless valve leakage monitoring system with acoustic emission sensor," *IEEE Transactions on Circuits and Systems-I: Regular Papers*, vol. 64, no. 11, pp. 2884–2893, 2017.
11. A. Collado and A. Georgiadis, "Conformal hybrid solar and electromagnetic (EM) energy harvesting rectenna," *IEEE Transactions on Circuits and Systems-I: Regular Papers*, vol. 60, no. 8, pp. 2225–2234, 2013.
12. M. Habibzadeh, M. Hassanalieragh, A. Ishikawa, T. Soyata, and G. Sharma, "Hybrid solar-wind energy harvesting for embedded applications: super capacitor-based system architectures and design trade-offs," *IEEE Circuits and Systems Magazine*, vol. 17, no. 4, pp. 29–63, 2017.

13. X. Gu, W. Liu, L. Guo, S. Hemour, F. Formosa, A. Badel, and K. Wu, "Hybridization of integrated microwave and mechanical power harvester," *IEEE Access*, vol. 6, pp. 13921–13930, 2018.
14. M. Aien, M.G. Khajeh, M. Rashidinejad, and M. Fotuhi-Firuzabad, "Probabilistic power flow of correlated hybrid wind-photovoltaic power systems," *IET Renewable Power Generation*, vol. 8, no. 6, pp. 649–658, 2014.
15. W.C. Brown, R.H. George, N.I. Heenan, and R.C. Wonson, "Microwave to dc converter," US Patent US3434678A, filed 05 May 1965 and issued 25 March 1969.
16. W.C. Brown, "The technology and application of free-space power transmission by microwave beam," *Proceedings of the IEEE*, vol. 62, no. 1, pp. 11–25, 1974.
17. F. Congedo, G. Monti, L. Tarricone, and V. Bella, "A 2.45-GHz vivaldi rectenna for the remote activation of an end device radio node," *IEEE Sensors Journal*, vol. 13, no. 9, pp. 3454–3461, 2013.
18. O.O. Conchubhair, K. Yang, P. McEvoy, and M.J. Ammann, "Amorphous silicon solar vivaldi antenna," *IEEE Antennas and Wireless Propagation Letters*, vol. 15, pp. 893–896, 2016.
19. A. Takacs, H. Aubert, S. Fredon, L. Despoisse, and H. Blondeaux, "Microwave power harvesting for satellite health monitoring," *IEEE Transactions on Microwave Theory and Techniques*, vol. 62, no. 4, pp. 1090–1098, 2014.
20. G.A. Vera, S.D. Nawale, Y. Duroc, and S. Tedjini, "Read range enhancement by harmonic energy harvesting in passive UHF RFID," *IEEE Microwave and Wireless Components Letters*, vol. 25, no. 9, pp. 627–629, 2015.
21. M. Siniscalchi, A. Pieruccioni, F. Vanzini, L. Reyes, and L. Barboni, "Schottky diode assessment for implementing a rectenna for radio-triggered wireless sensor networks," *IEEE Microwave and Wireless Components Letters*, vol. 27, no. 8, pp. 763–765, 2017.
22. A. Kiourti and K.S. Nikita, "A review of implantable patch antennas for biomedical telemetry: challenges and solutions," *IEEE Antennas and Propagation Magazine*, vol. 54, no. 3, pp. 210–228, 2012.
23. S. Hemour, Y. Zhao, C.H.P. Lorenz, D. Houssameddine, Y. Gui, C.M. Hu, and K. Wu, "Towards low- power high-efficiency RF and microwave energy harvesting," *IEEE Transactions on Microwave Theory and Techniques*, vol. 62, no. 4, pp. 965–976, 2014.
24. N.B. Carvalho, A. Georgiadis, A. Costanzo, H. Rogier, A. Collado, J.A. García, S. Lucyszyn, P. Mezzanotte, J. Kracek, D. Masotti, and A.J.S. Boaventura, "Wireless power transmission: R&D activities within Europe," *IEEE Transactions on Microwave Theory and Techniques*, vol. 62, no. 4, pp. 1031–1045, 2014.
25. K. Shafique, B.A. Khawaja, M.D. Khurram, S.M. Sibtain, Y. Siddiqui, M. Mustaqim, H.T. Chattha, and X. Yang, "Energy harvesting using a low-cost rectenna for internet of things (IoT) applications," *IEEE Access*, vol. 6, pp. 30932–30941, 2018.
26. S.H. ZainudDeen, H.A. ElAraby, and H.A.E.A. Malhat, "Energy harvesting enhancement of nanoantenna coupled to geometric diode using terhertz transmitarray," *Wireless Personal Communications*, vol. 107, pp. 159–168, 2019.
27. H. Jabbar, Y.S. Song, and T.T. Jeong, "RF energy harvesting system and circuits for charging of mobile devices," *IEEE Transactions on Consumer Electronics*, vol. 56, no. 1, pp. 247–253, February 2010.
28. M. Mrnka, P. Vasina, M. Kufa, V. Hebelka, and Z. Raida, "The RF energy harvesting antennas operating in commercially deployed frequency bands: a comparative study," *International Journal of Antennas and Propagation*, vol. 2016, Article ID 7379624, pp. 11, 2016.
29. W. Saeed, N. Shoaib, H.M. Cheema, and M.U. Khan, "RF energy harvesting for ubiquitous, zero power wireless sensors," *International Journal of Antennas and Propagation*, vol. 2018, Article ID 8903139, pp. 16, 2018.

30. L.K. Divakaran, D.D. Krishna, and Nasimuddin, "RF energy harvesting systems: an overview and design issues," *International Journal of RF and Microwave Computer-Aided Engineering*, vol. 29, no. 1, pp. 1–15, January 2019.
31. N. Tesla, "Apparatus for transmitting electrical energy," US patent number 1,119,732, 1914.
32. T.W.R. East, "A Self-steering Array for the SHARP Microwave-Powered Aircraft," *IEEE Transactions on Antennas and Propagation*, vol. 40, no. 12, pp. 1565–1567, 1992.
33. Z. Harouni, L. Cirio, L. Osman, A. Gharsallah, and O. Picon, "A dual circularly polarized 2.45-GHz rectenna for wireless power transmission," *IEEE Antennas and Wireless Propagation Letters*, vol. 10, pp. 306–309, April 2011.
34. N. Zainol, Z. Zakaria, M. Abu, and M.M. Yunus, "A 2.45 GHz harmonic suppression rectangular patch antenna with circular polarization for wireless power transfer application," *IETE Journal of Research*, vol. 64, pp. 310–316, 2018.
35. H. Sun, "An enhanced rectenna using differentially-fed rectifier for wireless power transmission," *IEEE Antennas and Wireless Propagation Letters*, vol. 15, pp. 32–35, April 2016.
36. S.T. Khang, D.J. Lee, I.J. Hwang, T.D. Yeo, and J.W. Yu, "Microwave power transfer with optimal number of rectenna arrays for midrange applications," *IEEE Antennas and Wireless Propagation Letters*, vol. 17, no. 1, pp. 155–159, 2018.
37. M. Wang, L. Yang, Y. Fan, M. Shen, Y. Li, and Y. Shi, "A compact omnidirectional dual-circinal rectenna for 2.45 GHz wireless power transfer," *International Journal of RF and Microwave Computer-Aided Engg*, vol. 29, pp. 1–7, 2019.
38. J.O. McSpadden, L. Fan, and K. Chang, "Design and experiments of a high-conversion efficiency 5.8 GHz rectenna," *IEEE Transactions on Microwave Theory and Techniques*, vol. 46, no. 912, pp. 2053–2060, 1998.
39. B. Strassner and K. Chang, "5.8-GHz circularly polarized rectifying antenna for wireless microwave power transmission," *IEEE Transactions on Microwave Theory and Techniques*, vol. 50, no. 8, pp. 1870–1876, 2002.
40. B. Strassner and K. Chang, "Highly efficient C-band circularly polarized rectifying antenna array for wireless microwave power transmission," *IEEE Transactions on Antennas and Propagation*, vol. 51, no. 6, pp. 1347–1356, 2003.
41. B. Strassner, and K. Chang, "5.8-GHz circularly polarized dual-rhombic-loop traveling-wave rectifying antenna for low power-density wireless power transmission applications," *IEEE Transactions on Microwave Theory and Techniques*, vol. 51, no. 5, pp. 1548–1553, May 2003.
42. J. Heikkinen and M. Kivikoski, "Low-profile circularly polarized rectifying antenna for wireless power transmission at 5.8 GHz," *IEEE Microwave and Wireless Components Letters*, vol. 14, no. 4, pp. 162–164, 2004.
43. Y. Yang, J. Li, L. Li, Y. Liu, B. Zhang, H. Zhu, and K. Huang, "A 5.8 GHz circularly polarized rectenna with harmonic suppression and rectenna array for wireless power transfer," *IEEE Antennas and Wireless Propagation Letters*, vol. 17, no. 7, pp. 1276–1280, July 2018.
44. Y.J. Ren and K. Chang, "New 5.8-GHz circularly polarized retrodirective rectenna arrays for wireless power transmission," *IEEE Transactions on Microwave Theory and Techniques*, vol. 54, no. 7, pp. 2970–2976, 2006.
45. Y.J. Ren, and K. Chang, "5.8-GHz circularly polarized dual-diode rectenna and rectenna array for microwave power transmission," *IEEE Transactions on Microwave Theory and Techniques*, vol. 54, no. 4, pp. 1495–1502, April 2006.
46. X.X Yang, C. Jiang, A.Z. Elsherbeni, F. Yang, and Y.Q. Wang, "A novel compact printed rectenna for data communication systems," *IEEE Transactions on Antennas and Propagation*, vol. 61, no. 5, pp. 2532–2539, May 2013.

47. R.H. Rasshofer, M.O. Thieme, and E.M. Biebl, "Circularly polarized millimeter-wave rectenna on silicon substrate," *IEEE Transactions on Microwave Theory and Techniques*, vol. 46, no. 5, pp. 715–718, 1998.
48. M. Ali, G. Yang, and R. Dougal, "A new circularly polarized rectenna for wireless power transmission and data communication," *IEEE Antennas and Wireless Propagation Letters*, vol. 4, pp. 205–208, April 2005.
49. S. Ladan, A.B. Guntupalli, and K. Wu, "A high-efficiency 24 GHz rectenna development towards millimeter-wave energy harvesting and wireless power transmission," *IEEE Transactions on Circuits and Systems-I: Regular Papers*, vol. 61, no. 12, pp. 3358–3366, December 2014.
50. C.A.D. Carlo, L.D. Donato, G.S. Mauro, R.L. Rosa, P. Livreri, and G. Sorbello, "A circularly polarized wideband high gain patch antenna for wireless power transfer," *Microwave and Optical Technology Letters*, vol. 60, pp. 620–625, 2018.
51. Y.J. Ren, M.Y. Li, and K. Chang, "35 GHz rectifying antenna for wireless power transmission," *Electronics Letters*, vol. 43, no. 11, pp. 1–2, May 2007.
52. T.W. Yoo and K. Chang, "Theoretical and experimental development of 10 and 35 GHz rectennas," *IEEE Transactions on Microwave Theory and Techniques*, vol. 40, no. 6, pp. 1259–1266, June 1992.
53. J.O. McSpadden, T. Yoo, and K. Chang, "Theoretical and experimental investigation of a rectenna element for microwave power transmission," *IEEE Transactions on Microwave Theory and Techniques*, vol. 40, no. 12, pp. 2359–2366, 1992.
54. Y.H. Suh and Kai Chang, "A high-efficiency dual-frequency rectenna for 2.45- and 5.8-GHz wireless power transmission," *IEEE Transactions on Microwave Theory and Techniques*, vol. 50, no. 7, pp. 1784–1789, July 2002.
55. T. Benyetho, J. Zbitou, L.E. Abdellaoui, H. Bennis, and A. Tribak, "A new fractal multiband antenna for wireless power transmission applications," *Active and Passive Electronic Components*, vol. 2018, Article ID 2084747, pp. 10, 2018.
56. N. Hassan, Z. Zakaria, W.Y. Sam, I.N.M. Hanapiah, A.N. Mohamad, A.F. Roslan, B.H. Ahmad, M.K. Ismail, and M.Z.A. Abd Aziz, "Design of dual-band microstrip patch antenna with right-angle triangular aperture slot for energy transfer application," *International Journal of RF and Microwave Computer-Aided Engineering*, vol. 29, no. 1, pp. e21666, January 2019.
57. Y. Shi, Y. Fan, Y. Li, L. Yang, and M. Wang, "An efficient broadband slotted rectenna for wireless power transfer at LTE band," *IEEE Transactions on Antennas and Propagation*, vol. 67, no. 2, pp. 814–822, Febuary 2019.
58. R. Andryczyk, P. Foldes, J. Chestek, and B.M. Kaupang, "Solar power satellites ground stations: The ground systems for microwave beaming from the SPS would require a rectifying antenna with over 13 billion elements," *Microwave Symposium Digest*, vol. 67, pp. 51–55, 1979.
59. R.V. Gelsthorpe and P.Q. Collins, "Increasing power input to a single solar power satellite rectenna by using a pair of satellites," *Electronics Letters*, vol. 16, no. 9, pp. 311–313, 1980.
60. T. Paing, J. Shin, R. Zane, and Z. Popovic, "Resistor emulation approach to low-power RF energy harvesting," *IEEE Transactions on Power Electronics*, vol. 23, no. 3, pp. 1494–1501, May 2008.
61. H. Sun, Y. Guo, M. He, and Z. Zhong, "Design of a high-efficiency 2.45-GHz rectenna for low-input-power energy harvesting," *IEEE Antennas and Wireless Propagation Letters*, vol. 11, pp. 929–932, August 2012.
62. C. Phongcharoenpanich and K. Boonying, "A 2.4-GHz dual polarized suspended square plate rectenna with inserted annular rectangular ring slot," *International Journal of RF and Microwave Computer-Aided Engineering*, vol. 26, no. 2, pp. 164–173, February 2016.

63. Y.S. Chen and J.W. You, "A scalable and multidirectional rectenna system for RF energy harvesting," *IEEE Transactions on Components, Packaging and Manufacturing Technology*, vol. 8, no. 12, pp. 2060–2072, December 2018.
64. Y. Shi, J. Jing, Y. Fan, L. Yang, and M. Wang, "Design of a novel compact and efficient rectenna for WiFi energy harvesting," *Progress in Electromagnetics Research C*, vol. 83, pp. 57–70, 2018.
65. Q. Awais, Y. Jin, H.T. Chattha, M. Jamil, H. Qiang, and B.A. Khawaja, "A compact rectenna system with high conversion efficiency for wireless energy harvesting," *IEEE Access*, vol. 6, pp. 35857–35866, 2018.
66. E.L. Chuma, Y. Iano, M.S. Costa, L.T. Manera, and L.L.B. Roger, "A compact-integrated reconfigurable rectenna array for RF power harvesting with a practical physical structure," *PIER M*, vol. 70, pp. 89–98, 2018.
67. Y. Shi, J. Jing, Y. Fan, L. Yang, J. Pang, M. Wang, "Efficient RF energy harvest with a novel broadband Vivaldi rectenna," *Microwave and Optical Technology Letters*, vol. 60, pp. 2420–2425, February 2018.
68. K. Celik and E. Kurt, "A novel meander line integrated E-shaped rectenna for energy harvesting applications," *International Journal of RF and Microwave Computer-Aided Engineering*, vol. 29, no. 1, pp. 1–10, January 2019.
69. W. Haboubi, H. Takhedmit, J.D.L.S. Luk, S.E. Adami, B. Allard, F. Costa, C. Vollaire, O. Picon, and L. Cirio, "An efficient dual-circularly polarized rectenna for RF energy harvesting in the 2.45 GHz ISM band," *Progress in Electromagnetics Research*, vol. 148, pp. 31–39, June 2014.
70. T. Mitani, S. Kawashima, and T. Nishimura, "Analysis of voltage doubler behavior of 2.45-GHz voltage doubler-type rectenna," *IEEE Transactions on Microwave Theory and Techniques*, vol. 65, no. 4, pp. 1051–1057, 2017.
71. F. Erkmen, T.S. Almoneef, and O.M. Ramahi, "Electromagnetic energy harvesting using full-wave rectification," *IEEE Transactions on Microwave Theory and Techniques*, vol. 65, no. 5, pp. 1843–1851, 2017.
72. Y. Chang, P. Zhang, and L. Wan, "Highly efficient differential rectenna for RF energy harvesting," *Microwave and Optical Technology Letters*, vol. 61, no. 12, pp. 2662–2668, December 2019.
73. M.A. Gozel, M. Kahriman, and O. Kasar, "Design of an efficiency-enhanced Greinacher rectifier operating in the GSM 1800 band by using rat-race coupler for RF energy harvesting applications," *International Journal of RF and Microwave Computer-Aided Engineering*, vol. 29, no. 1, pp. 1–8, January 2019.
74. X. Zhang, J. Grajal, J.L. Vazquez-Roy, U. Radhakrishna, X. Wang, W. Chern, L. Zhou, Y. Lin, P.C. Shen, X. Ji, and X. Ling, "Two-dimensional MoS2-enabled flexible rectenna for Wi-Fi-band wireless energy harvesting," *Nature*, vol. 566, pp. 368–383, F ebruary 2019.
75. U. Olgun, C.C. Chen, and J.L. Volakis, "Investigation of rectenna array configurations for enhanced RF power harvesting," *IEEE Antennas and Wireless Propagation Letters*, vol. 10, pp. 262–265, April 2011.
76. E.L. Chuma, YuzoIano, M.S. Costa, L.T. Manera, and L.L. Bravo Roger, "A compact integrated reconfigurable rectenna array for RF power harvesting with a practical physical structure," *PIER M*, vol. 70, pp. 89–98, 2018.
77. S. Ladan, N. Ghassemi, A. Ghiotto, and K. Wu, "Highly efficient compact rectenna for wireless energy harvesting application," *IEEE Microwave Magazine*, vol. 14, no. 1, pp. 117–122, February 2013.
78. S.D. Assimonis, V. Fusco, A. Georgiadis, and T. Samaras, "Efficient and sensitive electrically small rectenna for ultra-low power RF energy harvesting," *Scientific Reports*, vol. 8, pp. 1–13, 2018.

79. M. Palandoken and C. Gocen, "A modified Hilbert fractal resonator based rectenna design for GSM900 band RF energy harvesting applications," *International Journal of RF and Microwave Computer-Aided Engg*, vol. 29, pp. 1–8, 2019.
80. M. Zeng, A.S. Andrenko, X. Liu, Z. Li, and H.Z. Tan, "A compact fractal loop rectenna for RF energy harvesting," *IEEE Antennas and Wireless Propagation Letters*, vol. 16, pp. 2424–2427, 2017.
81. S. Ghosh, "Design and testing of rectifying antenna for RF energy scavenging in GSM 900 band," *International Journal of Computers and Applications*, vol. 39, no. 1, pp. 36–44, 2017.
82. H. Sun and W. Geyi, "A new rectenna using beamwidth-enhanced antenna array for RF power harvesting applications," *IEEE Antennas and Wireless Propagation Letters*, vol. 16, pp. 1451–1454, June 2017.
83. D. Kumar and K. Chaudhary, "Design of an improved differentially fed antenna array for RF energy harvesting," *IETE Journal of Research*, pp. 1–6, 2018.
84. N. Saranya and T. Kesavamurthy, "Design and performance analysis of broadband rectenna for an efficient RF energy harvesting application," *International Journal of RF and Microwave Computer-Aided Engineering*, vol. 29, no. 1, pp. 1–12, 2019.
85. H.P. Partal, MA. Belen, and S.Z. Partal, "Design and realization of an ultra-low power sensing RF energy harvesting module with its RF and DC sub-components," *International Journal of RF and Microwave Computer-Aiding Engineering*, vol. 29, no. 1, pp. 1–12, 2019.
86. K.T. Chandrasekaran, Nasimuddin, A. Alphones, and M.F. Karim, "Compact circularly polarized beam-switching wireless power transfer system for ambient energy harvesting applications" *International Journal of RF And Microwave Computer-Aiding Engineering*, vol. 29, no. 1, pp. 1–10, January 2019.
87. S.B. Vignesh, Nasimuddin, and A. Alphones, "Circularly polarized strips integrated microstrip antenna for energy harvesting applications," *Microwave and Optical Technology Letters*, vol. 58, no. 5, pp. 1044–1049, May 2016.
88. A. Mavaddat, S.H.M. Armaki, and A.R. Erfanian, "Millimeter wave energy harvesting using 4×4 microstrip patch antenna array," *IEEE Antennas and Wireless Propagation Letters*, vol. 14, pp. 515–518, 2015.
89. A.A. Masius, Y.C. Wong, and K.T. Lau, "Miniature high gain slot-fed rectangular dielectric resonator antenna for IoT RF energy harvesting," *International Journal of Electronics & Communication (AEÜ)*, vol. 85, pp. 39–46, February 2018.
90. H. Sun, Y.X. Guo, M. He, and Z. Zhong, "A dual-band rectenna using broadband yagi antenna array for ambient RF power harvesting," *IEEE Antennas and Wireless Propagation Letters*, vol. 12, pp. 918–921, 2013.
91. I. Adam, M.N.M. Yasin, H.A. Rahim, P.J. Soh, and M.F. Abdulmalek, "A compact dual-band rectenna for ambient RF energy harvesting," *Microwave and Optical Technology Letters*, vol. 60, pp. 2740–2748, 2018.
92. L. Zhu, J. Zhang, W. Han, L. Xu, and X. Bai, "A novel RF energy harvesting cube based on air dielectric antenna arrays," *International Journal of RF and Microwave Computer-Aiding Engineering*, vol. 29, no. 1, pp. 1–7, January 2019.
93. Z. Liu, Z. Zhong, and Y.X. Guo, "Enhanced dual-band ambient RF energy harvesting with ultra-wide power range," *IEEE Microwave and Wireless Components Letters*, vol. 25, no. 9, pp. 630–632, September 2015.
94. S. Ghosh and A. Chakrabarty, "Dual band circularly polarized monopole antenna design for RF energy harvesting," *IETE Journal of Research*, vol. 62, no. 1, pp. 9–16, 2016.
95. D.K. Ho, V.D. Ngo, I. Kharrat, T.P. Vuong, Q.C. Nguyen, and M.T. Le, "A Novel dual-band rectenna for ambient RF energy harvesting at GSM 900 MHz and 1800 MHz," *Advances in Science, Technology and Engineering Systems Journal*, vol. 2, no. 3, pp. 612–616, 2017.

96. M. Zeng, Z. Li, A.S. Andrenko, Y. Zeng, and H.Z. Tan, "A compact dual-band rectenna for GSM900 and GSM1800 energy harvesting," *International Journal of Antennas and Propagation*, vol. 2018, Article ID 4781465, pp. 9, 2018.
97. S. Shrestha, S.R. Lee, and D.Y. Choi, "A new fractal-based miniaturized dual band patch antenna for RF energy harvesting," *International Journal of Antennas and Propagation*, vol. 2014, Article ID 805052, pp. 9, 2014.
98. M. Aboualalaa, A.B.A. Rahman, A. Allam, H. Elsadek, and R.K. Pokharel, "Design of a dual-band microstrip antenna with enhanced gain for energy harvesting applications," *IEEE Antennas and Wireless Propagation Letters*, vol. 16, pp. 1622–1626, 2017.
99. F.S.M. Noor, Z. Zakaria, H. Lago, and M.A.M. Said, "Dual-band aperture-coupled rectenna for radio frequency energy harvesting," *International Journal of RF and Microwave Computer-Aiding Engineering*, vol. 29, no. 1, pp. 1–9, January 2019.
100. Z. Chen, M. Zeng, A.S. Andrenko, Y. Xu, and H.Z. Tan, "A dual-band high-gain quasi-Yagi antenna with split-ring resonators for radio frequency energy harvesting," *Microwave and Optical Technology Letters*, vol. 61, pp. 2174–81, 2019.
101. N. Singh, B.K. Kanaujia, M.T. Beg, Mainuddin, S. Kumar, and M.K. Khandelwal, "A dual band rectifying antenna for RF energy harvesting," *Journal of Computational Electronics*, vol. 17, pp. 1748–1755, 2018.
102. A.M. Jie, M.F. Karim, and K.T. Chandrasekaran, "A dual-band efficient circularly polarized rectenna for RF energy harvesting systems," *International Journal of RF and Microwave Computer-Aiding Engineering*, vol. 29, pp. 1–11, 2019.
103. O. Amjad, S.W. Munir, Ş.T. Imeci, and A.O. Ercan, "Design and implementation of dual band microstrip patch antenna for WLAN energy harvesting system," *ACES Journal*, vol. 33, no. 7, pp. 746–751, July 2018.
104. M. Wang, Y. Fan, L. Yang, Y. Li, J. Feng, and Y. Shi, "Compact dual-band rectenna for RF energy harvest based on a tree-like antenna," *IET Microwaves, Antennas & Propagation*, vol. 13, no. 9, pp. 1350–1357, 2019.
105. N. Singh, B.K. Kanaujia, M.T. Beg, Mainuddin, and S. Kumar, "A triple band circularly polarized rectenna for RF energy harvesting," *Electromagnetics*, vol. 39, no. 7, pp. 481–490, Mar. 2019.
106. S. Shen, C.Y. Chiu, and R.D. Murch, "A dual-port triple-band l-probe microstrip patch rectenna for ambient RF energy harvesting," *IEEE Antennas and Wireless Propagation Letters*, vol. 16, pp. 3071–3074. 2017.
107. S. Chandravanshi, S.S. Sarma, and M.J. Akhtar, "Design of triple band differential rectenna for RF energy harvesting," *IEEE Transactions on Antennas and Propagation*, vol. 66, no. 6, pp. 2716–2726, June 2018.
108. V. Kuhn, C. Lahuec, F. Seguin, and C. Person, "A multi-band stacked RF energy harvester with RF-to-DC efficiency upto 84%," *IEEE Transactions on Microwave Theory and Techniques*, vol. 63, no .5, pp. 1768–1778, May 2015.
109. S. Agrawal, M.S. Parihar, and P.N. Kondekar, "A quad-band antenna for multi-band radio frequency energy harvesting circuit," *International Journal of Electronics and Communication (AEÜ)*, vol. 85, pp. 99–107, 2018.
110. S. Ullah, C. Ruan, T. U. Haq, A. K. Fahad and J. Dai, "Design of High Efficiency Multiband Rectenna for RF Energy Harvesting," *2018 Progress in Electromagnetics Research Symposium (PIERS-Toyama)*, Toyama, 2018, pp. 2340–2343. doi: 10.23919/PIERS.2018.8598195
111. E. Karakaya, F. Bagci, S. Can, A.E. Yilmaz, and B. Akaoglu, "Four-band electromagnetic energy harvesting with a dual-layer metamaterial structure," *International Journal of RF and Microwave Computer-Aiding Engg*, vol. 29, pp. 1–7, 2019.
112. C. Song, Y. Huang, P. Carter, J. Zhou, S. Yuan, Q. Xu, and M. Kod, "A novel six-band dual CP rectenna using improved impedance matching technique for ambient RF

energy harvesting," *IEEE Transactions on Antennas and Propagation*, vol. 64, no. 7, pp. 3160–3171, 2016.
113. N. Singh, B.K. Kanaujia, M.T. Beg, Mainuddin, T. Khan, and S. Kumar, "A dual polarized multiband rectenna for RF energy harvesting," *International Journal of Electronics and Communication (AEÜ)*, vol. 93, pp. 123–131, September 2018.
114. N. Singh, B.K. Kanaujia, M.T. Beg, Mainuddin, S. Kumar, H.C. Choi, and K.W. Kim, "Low profile multiband rectenna for efficient energy harvesting at microwave frequencies," *International Journal of Electronics*, vol. 106, pp. 2056–2071, 2019.
115. M. Arrawatia, M.S. Baghini, and G. Kumar, "Broadband bent triangular omnidirectional antenna for RF energy harvesting," *IEEE Antennas and Wireless Propagation Letters*, vol. 15, pp. 36–39, 2016.
116. S. Agarwal, R.D. Gupta, M.S. Parihar, and P.N. Kondekar, "A wideband high gain dielectric resonator antenna for RF energy harvesting application," *International Journal of Electronics and Communications (AEU)*, vol. 78, pp. 24–31, August 2017.
117. H. Kumar, M. Arrawatia, and G. Kumar, "Broadband planar log-periodic dipole array antenna based RF-energy harvesting system," *IETE Journal of Research*, pp. 39–43, 2017.
118. S. Agrawal, M.S. Parihar, and P.N. Kondekar, "Broadband rectenna for radio frequency energy harvesting application," *IETE Journal of Research*, vol. 64, pp. 347–353, 2018.
119. H. Mahfoudi, M. Tellache, and H. Takhedmit, "A wideband rectifier array on dual-polarized differential-feed fractal slotted ground antenna for RF energy harvesting," *International Journal of RF and Microwave Computer-Aiding Engineering*, vol. 29, no. 8, pp. 1–7, August 2019.
120. M Mansour, X.L. Polozec, and H. Kanaya, "Enhanced broadband RF differential rectifier integrated with archimedean spiral antenna for wireless energy harvesting applications," *Sensors*, vol. 19, pp. 655, 2019.
121. C. Song, Y. Huang, J. Zhou, J. Zhang, S. Yuan, and P. Carter, "A high-efficiency broadband rectenna for ambient wireless energy harvesting," *IEEE Transactions on Antennas and Propagation*, vol. 63, no. 8, pp. 3486–3495, 2015.
122. C.H.P. Lorenz, S. Hemour, W. Li, Y. Xie, J. Gauthier, P. Fay, and K. Wu, "Breaking the efficiency barrier for ambient microwave power harvesting with hetero-junction backward tunnel diodes," *IEEE Transactions on Microwave Theory and Techniques*, vol. 63, no. 12, pp. 4544–4555, 2015.
123. B. Kapilevich, V. Shashkin, B. Litvak, G. Yemini, A. Etinger, D. Hardon, and Y. Pinhasi, "W-Band rectenna coupled with low-barrier mott diode," *IEEE Microwave and Wireless Components Letters*, vol. 26, no. 8, pp. 637–639, 2016.
124. Y.S. Chen and C.W. Chiu, "Maximum achievable power conversion efficiency obtained through an optimized rectenna structure for rf energy harvesting," *IEEE Transactions on Antennas and Propagation*, vol. 65, no. 5, pp. 2305–2317, 2017.
125. T. Li, K. Sawada, H. Ogai, and W. Si, "UHF-Band wireless power transfer system for structural health monitoring sensor network," *Smart Materials Research*, vol. 2013, Article ID 496492, pp. 7, 2013.
126. J.W. Zhang, K.Y. See, and T. Svimonishvili, "Printed decoupled dual-antenna array on-package for small wirelessly powered battery-less device," *IEEE Antennas and Wireless Propagation Letters*, vol. 13, pp. 923–926, 2014.
127. S.T. Khang, J.W. Yu, and W.S. Lee, "Compact folded dipole rectenna with RF-based energy harvesting for IoT smart sensors," *Electronics Letters*, vol. 51, no. 12, pp. 926–928, 2015.
128. A. Bakkali, J. Pelegri-Sebastia, T. Sogorb, V. Llario, and A. Bou-Escriva, "A dual-band antenna for rf energy harvesting systems in wireless sensor networks," *Journal of Sensors*, vol. 2016, Article ID 5725836, pp. 8, 2016.

129. A. M. AbdelTawab and A. Khattab, "Efficient multi-band energy Harvesting circuit for Wireless Sensor nodes," *2016 Fourth International Japan-Egypt Conference on Electronics, Communications and Computers (JEC-ECC)*, Cairo, 2016, pp. 75–78. doi: 10.1109/JEC-ECC.2016.7518971
130. L. Yang, Y.J. Zhou, C. Zhang, X.M. Yang, X.X. Yang, and C. Tan, "Compact multiband wireless energy harvesting based battery-free body area networks sensor for mobile healthcare," *IEEE Journal of Electromagnetics, RF, and Microwaves in Medicine and Biology*, vol. 2, no. 2, pp. 109–115, June 2018.
131. G. Monti, L. Corchia, and L. Tarricone, "UHF wearable rectenna on textile materials," *IEEE Transactions on Antennas and Propagation*, vol. 61, no. 7, pp. 3869–3873, 2013.
132. M.K. Hosain, A.Z. Kouzani, M.F. Samad, and S.J. Tye, "A miniature energy harvesting rectenna for operating a head-mountable deep brain stimulation device," *IEEE Access*, vol. 3, pp. 223–234, 2015.
133. A. Abdi and H. Aliakbarian, "A miniaturized UHF-band rectenna for power transmission to deep-body implantable devices," *Cardiovascular Devices and Systems*, vol. 7, pp. 1900311–1900321, 2019.
134. B.J. DeLong, A. Kiourti, and J.L. Volakis, "A radiating near-field patch rectenna for wireless power transfer to medical implants at 2.4 GHz," *IEEE Journal of Electromagnetics, RF and Microwaves in Medicine and Biology*, vol. 2, no. 1, pp. 64–69, 2018.
135. F.J. Huang, C.M. Lee, C.L. Chang, L.K. Chen, T.C. Yo, and C.H. Luo, "Rectenna application of miniaturized implantable antenna design for triple-band biotelemetry communication," *IEEE Transactions on Antennas and Propagation*, vol. 59, no. 7, pp. 2646–2653, July 2011.
136. O.M. Sanusi, F.A. Ghaffar, A. Shamim, M. Vaseem, Y. Wang, and L. Roy, "Development of a 2.45 GHz antenna for flexible compact radiation dosimeter tags," *IEEE Transactions On Antennas and Propagation*, vol. 67, no. 8, pp. 5063–5072, August 2019.
137. M. Haerinia and S. Noghanian, "A printed wearable dual-band antenna for wireless power transfer," *Sensors*, vol. 19, no. 1732, pp. 1–10, 2019.
138. T. Karacolak, A.Z. Hood, and E. Topsakal, "Design of a dual-band implantable antenna and development of skin mimicking gels for continuous glucose monitoring," *IEEE Transactions on Microwave Theory and Techniques*, vol. 56, no. 4, pp. 1001–1008, 2008.
139. A. Kiourti and K.S. Nikita, "Meandered versus spiral novel miniature pifas implanted in the human head: tuning and performance," *International Conference on Wireless Mobile Communication and Healthcare*, vol. 83, pp. 80–87, 2011.
140. A. Kiourti and K.S. Nikita, "Miniature scalp-implantable antennas for telemetry in the mics and ISM bands: design, safety considerations and link budget analysis," *IEEE Transactions on Antennas and Propagation*, vol. 60, no. 8, pp. 3568–3575, 2012.
141. J. Kim and Y.R. Samii, "Implanted antennas inside a human body: simulations, designs, and characterizations," *IEEE Transactions on Microwave Theory and Techniques*, vol. 52, no. 8, pp. 1934–1943, 2004.
142. J. Kim and Y.R. Samii, "SAR reduction of implanted planar inverted F antennas with non-uniform width radiator," *IEEE Antennas and Propagation Society International Symposium*, pp. 1091–1094, 2006.
143. J. Kim and Y.R. Samii, "Planar inverted-F antennas on implantable medical devices: meandered type versus spiral type," *Microwave and Optical Technology Letters*, vol. 48, no. 3, pp. 567–572, 2006.
144. H. Permana, Q. Fang, and I. Cosic, "3-layer implantable microstrip antenna optimised for retinal prosthesis system in MICS band," *International Symposium on Bioelectronics and Bio-information's*, pp. 65–68, 2011.
145. W.C. Liu, F.M. Yeh, and M. Ghavami, "Miniaturized implantable broadband antenna for biotelemetry communication," *Microwave and Optical Technology Letters*, vol. 50, no. 9, pp. 2407–2409, 2008.

Index

Note: **Bold** page numbers refer to tables and *italic* page numbers refer to figures.

Printed in the United States
By Bookmasters